# Introduction to PSpice®

**James G. Gottling**
The Ohio State University

**Circuit Analysis**
Second Edition

**David R. Cunningham**
University of Missouri-Rolla

**John A. Stuller**
University of Missouri-Rolla

JOHN WILEY & SONS, INC.
New York • Chichester • Brisbane • Toronto • Singapore

ISBN: 0 471 12489 3


Printed in the Unites States of America
10 9 8 7 6 5 4 3

# Table of Contents

# Preface

*Introduction to PSpice with Student Exercise Disk* supplements *Circuit Analysis*, a text for first courses in electric circuit analysis by Cunningham and Stuller. PSpice® is a personal computer version of SPICE, which is an acronym for *Simulation Program with Integrated Circuit Emphasis*. SPICE, written at the University of California, Berkeley in the early 1970s for mainframe computers, continues to evolve. Since MicroSim Corporation's introduction in 1984 of PSpice for use on personal computers and their placement of an Evaluation Version into the public domain in 1986, computer simulation of electric and electronic circuits has become an integral part of electrical engineering education.

Faculties at some schools prefer to postpone use of circuit simulation until the electronics course. However, we believe that using analysis, simulation, and laboratory experimentation provides students the most effective learning experience. Analysis develops computational skills and encourages students to focus on circuit approximations. Simulation leads students to explore parameter variations on circuit performance and to consider the effects of parasitic elements. Laboratory experimentation makes students deal with circuit reality and helps them to relate analysis and simulation with actual circuit behavior. Simulation and analysis provide different ways to understand how electrical circuits work. Each method reinforces the other. Using both provides greater insight.

MicroSim Corporation now refers to their collection of programs as *The Design Center®*. This set includes PSpice, Probe®, StmEd®, and Parts®. Programs StmEd and Parts help you to graphically create transient waveforms and to develop device models. This manual does not describe how to use StmEd and Parts. Probe is a post-processor program that you use to display PSpice analysis results and to measure circuit performance, using the results of PSpice simulation. Using Probe to examine PSpice output is like using a digitizing oscilloscope to measure circuit variables in an experiment. The recent implementation of *search* and *cursor movement* commands makes Probe an extremely effective tool for circuit analysis. Chapter 2 explains how to use these commands. Many of the problems in Chapter 3 and solutions in Chapter 4 illustrate use of these new commands.

Chapter 1 explains how to run the INSTALL program to install the Evaluation Version of PSpice, Probe, StmEd, and Parts onto the hard disk of your IBM or IBM-compatible computer. Also, this chapter guides you through two simple circuit analyses with PSpice and Probe, using either DOS commands or a Control-Shell program. The Control-Shell program, included with the DOS evaluation disks, allows you to run PSpice, Probe, StmEd, and Parts from a menu, as if they were one program. This chapter also explains how to install the Macintosh Evaluation Version and leads you through the two simple examples.

Chapter 2 describes the syntax of PSpice programs and how to use Probe. Twelve examples in this chapter show you how to make effective use of PSpice and Probe. Because this manual is for use with a linear circuit theory course, device and model statements for semiconductor devices are not given.

Chapter 3 contains 72 PSpice problems, which relate to examples, exercises, or problems in *Circuit Analysis* by Cunningham and Stuller. The *Student Exercise Disk* has circuit netlists for each of these problems that you can complete as exercises. The disk also contains the complete form of the netlist. The names of files on the Student Exercise Disk relate to the problem numbers. For example, the netlist exercise file for Problem 5.2 is "PR5_02.CIR" and the solution for this netlist is file "PR5_02s.CIR."

Chapter 4 gives solutions for five of the Chapter 4 problems. We hope that these illustrate effective use of PSpice and Probe and also demonstrate how to relate simulation and analysis effectively.

Appendix A provides a summary of PSpice and Probe commands for quick reference.

The EVALUATION disk that comes with *Introduction to PSpice* is a 1440-kbyte IBM-DOS disk that contains the text files for all of the example netlists. Macintosh users can convert this disk to Macintosh format using the Macintosh system program *Apple File Exchange*. This program is in the *Apple File Exchange* Folder within the *Apple Utilities* folder on one of your system software disks. The system disk where you will find *Apple File Exchange* depends on your specific Apple computer, but often is within the *Tidbits* disk. When you use *Apple File Exchange*, select **Text translation** from the **DOS-to-Mac** menu. Lock the original disk to protect it from viruses and to keep the example files in their original form. As a precaution against disk failure, it is good practice to copy the original disk and work from the copy, keeping the original in a safe place. See Chapter 1 for more details about DOS-to-Macintosh file conversion.

Students and colleagues at The Ohio State University provided much of the motivation for writing this book, as we sought to make computer simulation a significant feature in the teaching of circuit theory and electronics. Ms. Sunny Carlson at the MicroSim Corporation was very helpful throughout this project, keeping me up-to-date on successive versions of *The Design Center*. My sponsoring editor, William E. Hoffman, provided encouragement and advice throughout this project. We are grateful to Ms. Linda Licklider Smith, Jennifer Roderick, and others of the Houghton Mifflin editorial and production staff for their contributions to the production of this book.

# Chapter 1 Installing and Running PSpice and Probe

This chapter explains how to install and run PSpice and Probe[1] on an IBM or IBM-compatible computer or on a Macintosh computer. PSpice and Probe are two of the four programs in The Design Center set of four programs. PSpice analyzes circuits that you describe with netlists. A netlist is a text file description of a circuit and analysis options that you want PSpice to do. PSpice produces an output file that gives the results of the circuit analysis. If a PROBE command statement is in the netlist, then PSpice creates a second output file that Probe uses as its input file. With Probe, you can select, plot, and analyze circuit variables or measures of circuit performance. The last two programs in The Design Center set, StmEd and Parts, let you graphically specify a transient waveform for use in a netlist and describe a PSpice device model using data-sheet information. This manual does not describe how to run and use StmEd and Parts.

## 1.1 How to Install PSpice and Probe

This section describes how to install PSpice and Probe on either an IBM or IBM-compatible computer or on a Macintosh computer.

## 1.1.1 DOS Installation

The DOS evaluation version 6.0 of The Design Center comes on two high-density disks and requires your computer to have a coprocessor. Lock these disks to avoid the possibility of changing them by mistake and to prevent virus damage to them. Copy your original disks, work from the copies, and keep your originals in a safe place. To install The Design Center programs, run the INSTALL program by inserting disk 1 into drive A: and use the DOS command

        A:INSTALL

If you have 1024k of extended memory, INSTALL gives you the option to install the digital as well as the analog simulation part of PSpice. Without 1024k of extended memory, your choices are to install the analog-only version or quit. INSTALL asks you to confirm that the C: drive is the target of the installation and suggests creation of a subdirectory named MSEVAL60 on the C: drive for storage of The Design Center files. Normally you should accept both of these defaults by typing either the return key <Return> or the enter key <Enter>. Now the installation begins, and you see a dialog that shows the transfer of the various programs and files to your subdirectory. After installation, INSTALL asks if you wish to modify your AUTOEXEC.BAT file to include C:\MSEVAL60 in the PATH statement and set PSPICELIB=C:\MSEVAL60. These changes cause PSpice to search MSEVAL60 for The Design Center files and the standard device library. Finally, INSTALL asks if you wish to have the CONFIG.SYS file include BUFFERS=20 and FILES=20. Type either a lower case or upper case letter Y to let INSTALL handle these matters for you automatically, since these changes enable the PSpice and Probe programs to run. Finally, INSTALL offers to run SETUPDEV, which is in the MSEVAL60 subdirectory. This program,

---

[1] PSpice and Probe are registered trademarks of the MicroSim Corporation.

which you can run now or later, lets you select the monitor type and printer ports that PSpice, Probe, StmEd, and Parts use.

## 1.1.2 Macintosh Installation

The Macintosh evaluation version 6.0 of The Design Center comes on two 1440-kilobyte disks. Lock your original disks to avoid the possibility of changing them by mistake and to prevent virus damage to them. Make working copies of your original disks, lock and work from these copies, and keep your originals in a safe place. To install the set of four programs from these disks, create a new folder on your hard disk using **New Folder** (⌘-N) on the **File** menu. Give this folder any name that you wish, such as PSpice ƒ. If you have a virus protection program running, turn it off before unstuffing the PSpice applications. Before disabling the virus protection program, it is a good idea to check to see that your PSpice working disks are free of viruses. Now, select the Part1.PAK self-unstuffing file on disk 1 by clicking on its icon, and choose **Open** (⌘-O) on the **File** menu or simply double click on Part1.PAK. In the dialog window that appears, open your PSpice folder as the target for the installation and click the **Save** button to have Part1.PAK transfer all of its contents into your PSpice folder. Notice that the dialog allows you to create a folder, in case you had not already done so. Repeat this process with the Part2.PAK self-unstuffing file, which also is on disk 1, and with the Part3.PAK self-unstuffing file, which is on disk 2. Select your PSpice folder as the target for all three of these self-unstuffing files. The files that these self-unstuffing files create in your PSpice folder are

Part1.PAK:    PSpice

Part2.PAK:    Probe

Part3.PAK:    StmEd, STMED.HLP, Parts, PARTS.HLP, EVAL.LIB, Example1.CIR, Example1.OUT, Example1.DAT, EVALPWRS.CIR, EVALPWRS.OUT, EVALPWRS.DAT

If you have little available hard-drive memory, you can start using PSpice without installing any of the files from Part3.PAK.

## 1.2 The Design Center Files

PSpice uses an ASCII text input file. This file describes the circuit and tells PSpice which circuit analyses to perform. By convention, PSpice expects this file to have a name with a CIR suffix.[2] After processing this input file, PSpice creates an output file having the same name as the CIR file, but with an OUT suffix. Also, if the input CIR file includes a PROBE statement, then PSpice creates a second output file with the same name as the CIR file with a DAT suffix for use by Probe.

---

[2] Attach the CIR suffix to the file name "myCircuit" with a period, as in "myCircuit.CIR."

As you use PSpice to analyze a variety of circuits, you may accumulate many files. To maintain order on the hard disk of the machine that you use or to keep your files separate from those of other users of PSpice, you may wish either to create data subdirectories or folders on your hard disk or store your files on floppy disks. With a Macintosh computer, select the appropriate disk or folder when naming the file. With an IBM or IBM-compatible computer you may need to create a path between either the data subdirectory or the floppy disk and the MSEVAL60 subdirectory if you did not let INSTALL do this for you. The following sections  describe this process.

### 1.2.1 Using a Floppy Disk for Data

If you wish to keep your files on a floppy disk in drive A: and the PSpice program is in the C: hard disk within the MSEVAL60 subdirectory, you can load and run PSpice using drive A: as the current directory by establishing the proper DOS path. With the computer on and showing the DOS prompt C:\>, insert a formatted disk into drive A: and type

A: <Return>

To link the program files in the MSEVAL60 subdirectory with the data files on the floppy disk, at the A:\> prompt type

PATH=C:\MSEVAL60;C:\DOS <Return>

Then from the A:\ prompt run the PSpice Control Shell program or run PSpice in the batch mode. You do not need to type the PATH statement above if INSTALL modifies the PATH statement in your AUTOEXEC.BAT file during installation.

### 1.2.2 Using a Data Subdirectory on the Hard Disk

If you wish to keep your files in a subdirectory that you want to name SP_DATA on the hard disk C:, first create your subdirectory after the DOS prompt C:\> by typing

MD SP_DATA <Return>

Next type

CD SP_DATA <Return>

To link the program files in the MSEVAL60 directory with the data files in the SP_DATA subdirectory, at the C:\SP_DATA> prompt type

PATH=C:\MSEVAL60;C:\DOS <Return>

Then from the C:\SP_DATA> prompt run the Control Shell program or run PSpice in the batch mode. If INSTALL modifies the PATH in your AUTOEXEC.BAT  file for you during installation, then you do not need to change the PATH.

### 1.3 PSpice  Operation

Circuit analysis with PSpice involves the following steps:

> • Use a text editor to write a description of the circuit and state the analyses that you want PSpice to do. Save this file in ASCII format and

give the file a name using the suffix CIR. Although the CIR suffix is unnecessary with a Macintosh computer, using this suffix shows that the text file is a PSpice input file. Exit the text editor.

• Run PSpice with the text file as the input file. PSpice creates a file having output information that has the same name as your source file, but replaces the suffix CIR with OUT. Also, if a PROBE statement occurs in your input file, PSpice creates a second output file having your source file name and a DAT suffix. This file passes information to Probe. If you use an IBM PC and the CIR file has a PROBE statement, then after PSpice finishes, Probe runs automatically.

• Examine and interpret the OUT file with your text editor or run Probe using the DAT file as input to plot circuit variables.

Using two simple circuit examples, the next three sections show how to run PSpice and Probe using either DOS or the Control Shell on a PC or PC-compatible computer or on a Macintosh computer.

### 1.3.1 Running PSpice using DOS with an IBM Machine

To run PSpice directly from DOS, first create a netlist using any word processing program. You can use the DOS editor if you wish. Save this netlist as an ASCII file either in C:\MSEVAL60, another subdirectory, or onto a floppy disk. Next, either change the directory to C:\MSEVAL60 or follow either of the procedures in Sections 1.2.1 or 1.2.2. Then, at the DOS prompt, type a line having the form

SIM <Input_file> [<Output_file> [<Probe_file>]] <Return>

When you type this line replace <Input_file> with the name of your netlist. Replace <Output_file> with the name that you want to give to the output file and <Probe_file> with the name that you want the Probe file to have. The square brackets shown above, which you do not type, indicate that the last two file names are optional. You can omit the suffixes CIR, OUT, and DAT that PSpice associates with netlists, PSpice output files, and Probe input files, because PSpice assumes that the first file you name is a CIR file, the second is an OUT file, and the third is a DAT file. If you omit the OUT and DAT files, PSpice assumes that the OUT and DAT files have the same name as your input file. For example, if your input file has the name MyCircuit.CIR, any of the following statements after the DOS prompt run PSpice without further response by you.

SIM MyCircuit.CIR MyCircuit.OUT MyCircuit.DAT<Return>
SIM MyCircuit.CIR <Return>
SIM MyCircuit <Return>

Replace SIM with PSPICE if you want to run PSpice without automatically invoking Probe.

To see how to run PSpice, follow the instructions given next. With your PC running and showing the prompt C:\>, lock the DOS-format Examples Disk and insert it into drive A:. If you do not already have an SP_DATA subdirectory, after the DOS prompt C:\> type

```
 File  Edit  Search  Options                                    Help
┌──────────────────────────── EX_1_1.CIR ─────────────────────────────┐
EX_1_1 - A simple circuit
VS 1 0 1
R1 1 2 1K
R2 2 0 2K
.OP
.END_
```

FIGURE 1.1 The EX_1_1 Netlist

     MD  SP_DATA <Return>

Now, copy all of the example files from the Examples Disk onto your SP_DATA subdirectory and change to the SP_DATA subdirectory by typing

     COPY A:*.*  C:\SP_DATA <Return>
     CD  SP_DATA <Return>

     Look at the directory listing of subdirectory SP_DATA to see that a CIR file EX_1_1.CIR exists. To use the DOS editor to view EX_1_1.CIR, as shown in Fig. 1.1, type

     EDIT EX_1_1.CIR <Return>

This file, which we call a *netlist*, describes the circuit shown in Fig. 1.2. The circuit is a series connection of a 1-V DC voltage source $V_S$ with a 1-k$\Omega$

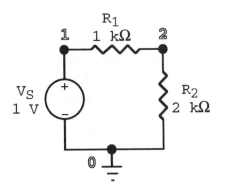

FIGURE 1.2 Circuit for Example 1.1

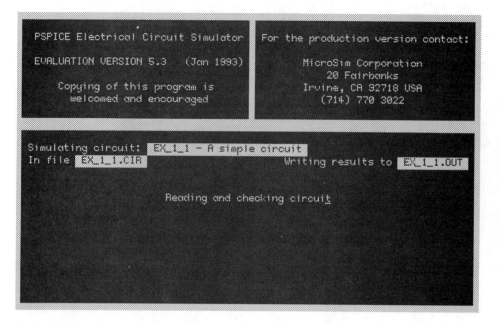

FIGURE 1.3 The IBM PSpice Banner[3]

resistance $R_1$ and a 2-k$\Omega$ resistance $R_2$. The netlist begins with a mandatory comment line that describes the circuit and ends with an END statement. The netlist shows that source $V_S$ connects between nodes 1 and 0, resistance $R_1$ connects between nodes 1 and 2, and resistance $R_2$ connects between nodes 2 and 0. The positive node of source $V_S$ is node 1. Although the order of the nodes for $R_1$ and $R_2$ has no significance here, SPICE assumes that the positive current direction is from the first to the second node. The OP statement specifies that the DC node voltages are to appear in the output file, together with the DC currents in any voltage sources and the total DC circuit power dissipation. SPICE considers upper- and lowercase characters to be the same. The period (.) at the beginning of both the OP and the END statement shows that these lines are command lines.

To run PSpice, quit the editor to return to DOS and type

      PSPICE EX_1_1.CIR <Return>

You will see the PSpice banner shown in Fig. 1.3 on your monitor screen. When PSpice finishes analyzing the circuit that netlist EX_1_1.CIR describes, PSpice closes and you will see a DOS prompt. Examine the SP_DATA directory to see that output file EX_1_1.OUT exists. Use the DOS editor to view EX_1_1.OUT to see the screen images shown in Figs. 1.4 and 1.5. These screen images show that the voltage at node 2 is 0.6667 V, the voltage-source current is

---

[3] DOS displays here use PSpice Version 5.3. PSpice Version 6.0 displays are different in some respects.

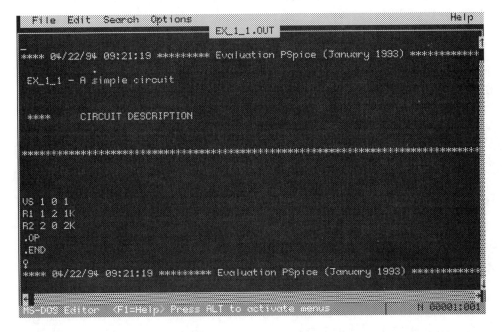

FIGURE 1.4 Start of the Output File

```
 File  Edit  Search  Options                            Help
                        EX_1_1.OUT
_
****  04/22/94  09:21:19 ********* Evaluation PSpice (January 1993) ***********

 EX_1_1 - A simple circuit


 ****      CIRCUIT DESCRIPTION



 ****************************************************************************




 VS 1 0 1
 R1 1 2 1K
 R2 2 0 2K
 .OP
 .END
 ¢
 ****  04/22/94  09:21:19 ********* Evaluation PSpice (January 1993) ***********

 MS-DOS Editor   <F1=Help>  Press ALT to activate menus        N 00001:001
```

```
 File  Edit  Search  Options                            Help
                        EX_1_1.OUT
_
 ****************************************************************************



 NODE   VOLTAGE   NODE   VOLTAGE   NODE   VOLTAGE   NODE   VOLTAGE

 (    1)   1.0000  (    2)    .6667




    VOLTAGE SOURCE CURRENTS
    NAME         CURRENT

    VS          -3.333E-04

    TOTAL POWER DISSIPATION   3.33E-04   WATTS

 MS-DOS Editor   <F1=Help>  Press ALT to activate menus        N 00027:001
```

FIGURE 1.5 The Rest of the Output File

-0.3333 mA from node 1 to 0, and that the total DC dissipation is 0.333 mW. Of course these values agree with values that you easily can calculate using the voltage divider theorem, the series equivalent resistance, Ohm's law, and the resistance power expression.

### 1.3.2 Running PSpice on a DOS Machine with the Control Shell

During installation of PSpice, INSTALL creates the Control Shell program in the MSEVAL60 subdirectory. You create circuit files, operate PSpice, Probe, StmEd, and Parts entirely from within this shell as if you were running a single program. The shell also includes a simple word processor program to create circuit files and view output files. To run the Control Shell program, either change the directory to MSEVAL60 or follow either of the procedures in Sections 1.2.1 or 1.2.2. Then, after the DOS prompt type

    PS <Return>

However, if you have a CGA adapter with a monochrome monitor or a Hercules monochrome graphics adapter, instead type

    PS -M <Return>

To follow the text now on your DOS computer, if you did not copy the example files from the Examples disk while reading Section 1.3.1 do so now and change to the SP_DATA subdirectory. Then type

    PS <Return>

The **Control Shell** menu screen shown in Fig. 1.6 appears on your monitor

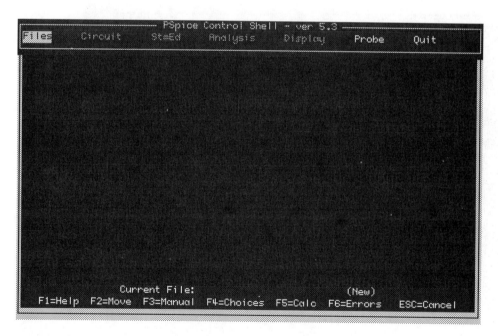

FIGURE 1.6 Control Shell Menu

screen. The highlighting of the **Files** menu shows that this menu is the current selection. **Probe** and **Quit** appear with normal intensity and are available for selection. **Circuit**, **StmEd**, **Analysis**, and **Display** appear with half intensity and are unavailable for selection now. You change the current selection using either the left- and right-arrow keys or by moving the mouse. By typing a menu's first letter, using either lower- or uppercase characters, you can both select a menu and open it.

Open a current selection by typing either <Enter>, <Return>, or the selection's first letter, using either an upper- or lowercase character. Open the **File** menu now to display the items shown in Fig. 1.7. The current item in the **File** menu is **Current File...** . You can change the current item using the up- and down-arrow keys or by mouse movement. Select a current item by typing <Enter> or <Return> or change the current item and select the new item by typing its first letter. Select **Current File...** now by typing <Return>. The dialog shown in Fig. 1.8 will appear. To the right of **Circuit File Name:** type the input file name EX_1_1 as shown here, leaving the cursor just to the right of the file name. This CIR file exists in the SP_DATA subdirectory, so the Control Shell can find it. Type the enter key <Enter> to return to the File menu and type character E to select **Edit** to see the EX_1_1 file, as shown in Fig. 1.9. Type <Escape> to leave the editor and <Escape> again to close the File menu. Type character A to select and open the **Analysis** menu, as shown in Fig. 1.10. Then run PSpice by typing <Return>. The PSpice banner shown in Fig. 1.3 now appears. Although the output file is EX_1_1.OUT, the source file has a scratch name that the control shell creates. However, you know that

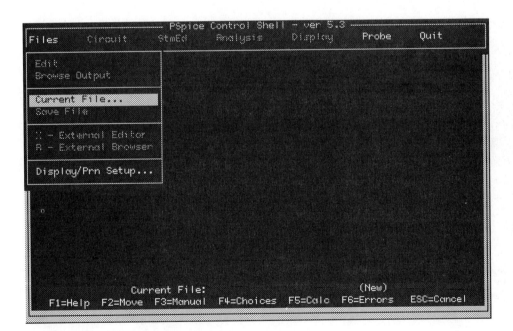

FIGURE 1.7 File Menu Items

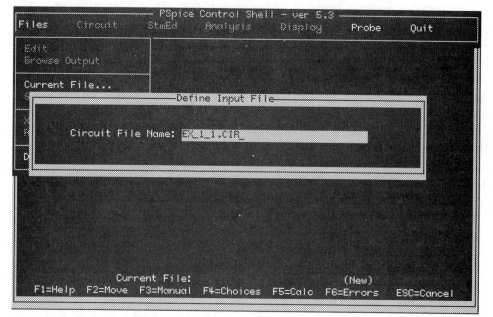

FIGURE 1.8 Dialog for Selection of Current File

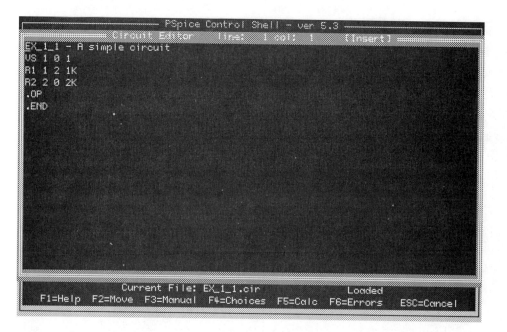

FIGURE 1.9 Viewing EX_1_1.CIR with Edit in the Control Shell

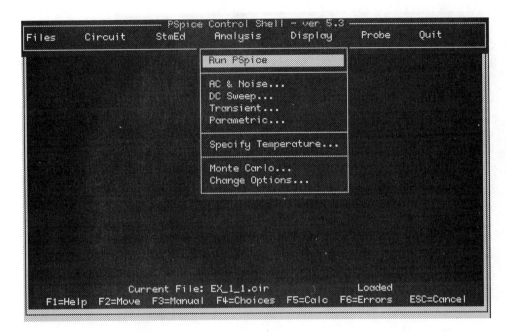

FIGURE 1.10 Select Run PSpice

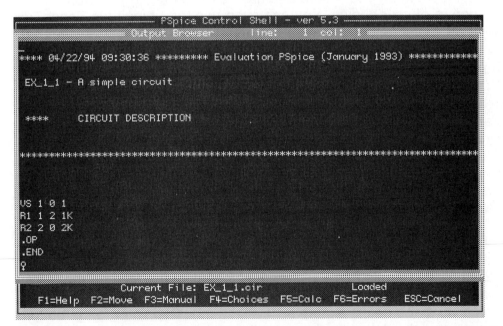

FIGURE 1.11 Browsing the Output File with Edit in the Control Shell

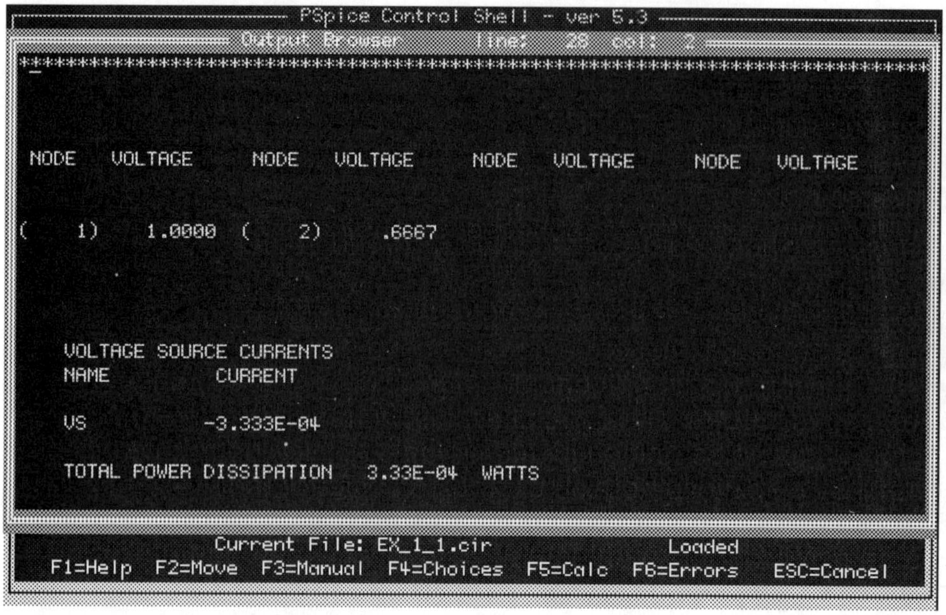

FIGURE 1.12 The Rest of the File

EX_1_1.CIR is the current file. Escape from the **Analysis** menu by typing <Escape>, select the **File** menu, and choose **Browse Output** by typing character B to see the output file EX_1_1.OUT in Fig. 1.11. Type <Page Down> to see the rest of the output file in Fig. 1.12.

### 1.3.3 Running PSpice on a Macintosh

With a Macintosh computer, you create and save a netlist using any word processing program of your choice. You can even use the ubiquitous TeachText program. However, be sure that you save the netlist in ASCII or text-only format.

Run PSpice by selecting the PSpice application and choosing **Open** (⌘-O) from the **File** menu, or simply double click on the application. A dialog then prompts you to identify the input file and another dialog prompts you to name the output file, suggesting your input file name with the OUT suffix as the default output file name. If your computer has sufficient random access memory (RAM), your word processing program and PSpice can run at the same time when you use either Multifinder with a system prior to 7.0 or system 7.0 or higher.

To see how this process works, copy the sample netlist files from the Examples disk to your Macintosh hard disk. First, lock the DOS Examples Disk, and create a new folder in your PSpice folder, naming it SP_DATA. Next, open the program Apple File Exchange. You can find this DOS-to-Macintosh translation program on your Macintosh system disks in the Apple File Exchange Folder. Open Apple File Exchange, and insert the DOS Examples disk into your floppy-disk drive. Within Apple File Exchange, select all of the files on the Examples disk by holding the shift key as you drag the mouse pointer through the file names of all of the netlist examples. Pull the mouse pointer to the bottom of the list box to scroll the selection through all of the files. From the **MS-DOS**

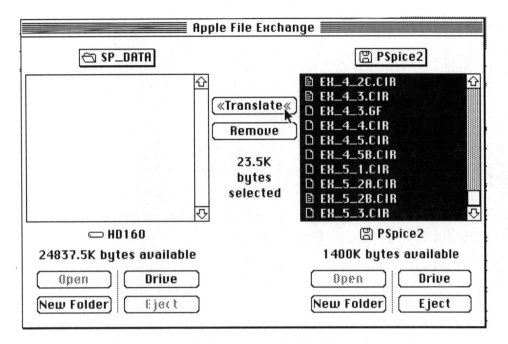

FIGURE 1.13 Apple File Exchange Dialog Window

**to Mac** menu deselect the **DCA-RFT to MacWrite** item and select **Text Translation...**, using all of the default items in the dialog. When the Apple File Exchange dialog appears as in Fig. 1.13, click on the **Translate** button to transfer all of the DOS files to your SP_DATA Macintosh folder.

　　With the example files now available in the SP_DATA folder, open PSpice by selecting the application and choosing **Open** (⌘-O) from the **File** menu or double click the PSpice application. PSpice opens a window that tells you to open files for input and output. Choose **Open** (⌘-O) from the PSpice **File** menu. Then select and open the SP_DATA folder either by clicking on

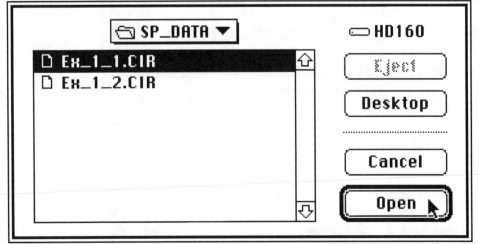

FIGURE 1.14 Input File Selection Dialog Window

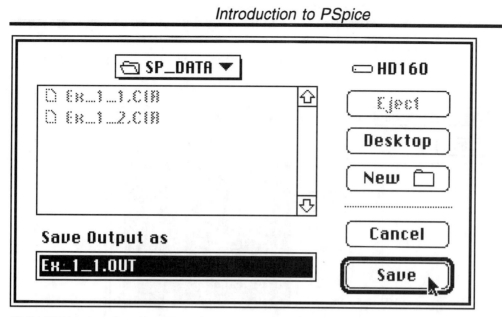

FIGURE 1.15 Output File Selection Dialog Window

SP_DATA in the list and typing <Return> or clicking on the **Open** button or by double clicking on SP_DATA in the list box. Next, open EX_1_1.CIR either by selecting EX_1_1.CIR by clicking on its name in the list and clicking the **Open** button or type <Return>, as shown in Fig. 1.14, or double click EX_1_1.CIR. After you select the circuit file, PSpice opens the dialog window shown in Fig. 1.15 to let you choose the output file name, suggesting EX_1_1.OUT as the default name. If this name is satisfactory, click on the **Open** button or type <Return> to accept the default name and the SP_DATA folder as the storage location. Now the Macintosh version of the PSpice banner shown in Fig. 1.16

```
▦□▤▦▦▦▦ Evaluation PSpice 6.0 from MicroSim Corp. ▦▦▦▦▦□▦
  ┌─────────────────────────────────┬────────────────────────────────┐
  │ PSPICE Electrical Circuit Simulator │    Macintosh II PSpice          │
  │ Evaluation Version 6.0 (Jan 1994)   │                                 │
  │                                     │ For production version, contact: │
  │     Copying this program is         │     MicroSim Corporation        │
  │     Welcomed and encouraged         │       (714) 770-3022            │
  └─────────────────────────────────┴────────────────────────────────┘

  Simulating circuit: Example 1.1 - A simple circuit
  In file Ex_1_1.CIR                    Writing results to Ex_1_1.OUT

               Calculating bias point
               Bias point calculated

                            ▸
```

FIGURE 1.16 Macintosh PSpice Banner

```
**** 04/15/94 19:33:47 ********** Evaluation PSpice (July 1993)

 Example 1.1 - A simple circuit

  ****       CIRCUIT DESCRIPTION

*****************************************************************
VS 1 0 1
R1 1 2 1k
R2 2 0 2k
.OP
.END
**** 04/15/94 19:33:47 ********** Evaluation PSpice (July 1993)

 Example 1.1 - A simple circuit

  ****   SMALL SIGNAL BIAS SOLUTION    TEMPERATURE =    27.000 DEG C

*****************************************************************
 NODE    VOLTAGE    NODE    VOLTAGE
( 1)     1.0000    ( 2)     .6667

    VOLTAGE SOURCE CURRENTS
    NAME           CURRENT
    VS             -3.333E-04
    TOTAL POWER DISSIPATION   3.33E-04   WATTS
```

FIGURE 1.17 EX_1_1.OUT File

appears. When PSpice finishes analyzing EX_1_1.CIR, the original PSpice window reappears. If you have sufficient RAM and are running with Multifinder or System 7.0 or higher you can let PSpice continue to run. If not, shut PSpice down either by clicking in the close box in the top left corner of the PSpice window or by selecting **Quit** (⌘-**Q**) from the PSpice **File** menu. Now run your word-processing program, open the output file EX_1_1.OUT, and examine and interpret the results. These will appear as shown in Fig. 1.17.

### 1.4 Probe  Operation

To start Probe from DOS on an IBM or IBM-compatible PC, type

      Probe [<Probe_file>] <Return>

You replace <Probe_file> with the name of the DAT file that you want Probe to use. From the Control Shell with **Auto-run** active, Probe runs automatically after PSpice successfully processes a netlist, even if a PROBE statement does not appear in the netlist. On an IBM PC, Probe runs automatically after PSpice closes when a PROBE statement appears in the CIR file.

      Using a Macintosh, start Probe in the same way as you would start any application. Either click on Probe's icon and select **Open** (⌘-**O**) from the **File** menu or double click Probe's icon. Probe begins by displaying a dialog window that tells you to open a file. Use Probe's **File** menu to select **Open** or type ⌘-**O** to see the file-selection dialog. This dialog makes available to you only DAT files, so you can select only a file that Probe can use. In the dialog window, click

```
Example 1.2 - Another simple circuit
Vs 1 0
Is 0 2 1m
R1 1 2 1k
R2 2 0 2k
.DC LIN Vs -5 5 1
.PRINT DC V(1,2) V(2)
.PROBE
.END
```

a) Netlist

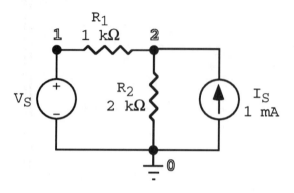

b) Circuit

FIGURE 1.18 EX_1_2.CIR Netlist and Circuit

on the file that you want to use, and either click the **Open** button or type
<Return>. Alternately, just double click the file in the dialog window. You also
can start Probe on a Macintosh by opening a DAT file. In this case, Probe omits
the file-selection process. With Probe running, a **Section Selection** menu
appears if your CIR file has more than one analysis. With only one analysis
command in the netlist, Probe goes directly to an Analog/Digital menu. To see
how to use this Analog/Digital menu, copy file EX_1_2.CIR from the Exercise
disk to your SP_DATA subdirectory or folder and run PSpice to create the
Probe input file EX_1_2.DAT.

Figure 1.18a shows the netlist EX_1_2.CIR, and the circuit that this
netlist describes appears in Fig. 1.18b. Of course, with $I_S$ set to zero this circuit
is the same as the EX_1_1 circuit. The netlist has no value for $V_S$, so the value
defaults to 0. This value affects the operating point values that PSpice includes
in the OUT file if the OP command appears. Since the OP command does not
appear in this netlist, the value of $V_S$ in the device statement has no effect on
the output file. The DC statement causes $V_S$ to change linearly (LIN) from -5 V
to 5 V in 1-V steps. The PRINT statement makes PSpice list the node voltage
difference $V_1 - V_2$ and the node voltage $V_2$ as a function of the source voltage
$V_S$. The PROBE statement in EX_1_2.CIR causes PSpice to create the file
EX_1_1.DAT. Figure 1.19 shows the output file after some editing.

If PSpice runs from either the Control Shell or from DOS using a PC, then
Probe opens automatically after PSpice closes, using the file EX_1_2.DAT. If
you use a Macintosh, open Probe after PSpice finishes and select EX_1_2.DAT
as the input file, or double click the EX_1_2.DAT file in the SP_DATA folder.

```
Example 1.2 - Another simple circuit
Vs 1 0
Is 0 2 1m
R1 1 2 1k
R2 2 0 2k
.DC LIN Vs -5 5 1
.PRINT DC V(1,2) V(2)
.PROBE
.END

 ****   DC TRANSFER CURVES          TEMPERATURE =    27.000 DEG C
  Vs          V(1,2)        V(2)
  -5.000E+00  -2.333E+00   -2.667E+00
  -4.000E+00  -2.000E+00   -2.000E+00
  -3.000E+00  -1.667E+00   -1.333E+00
  -2.000E+00  -1.333E+00   -6.667E-01
  -1.000E+00  -1.000E+00    0.000E+00
   0.000E+00  -6.667E-01    6.667E-01
   1.000E+00  -3.333E-01    1.333E+00
   2.000E+00   1.333E-09    2.000E+00
   3.000E+00   3.333E-01    2.667E+00
   4.000E+00   6.667E-01    3.333E+00
   5.000E+00   1.000E+00    4.000E+00
```

FIGURE 1.19 EX_1_2.OUT File After Editing

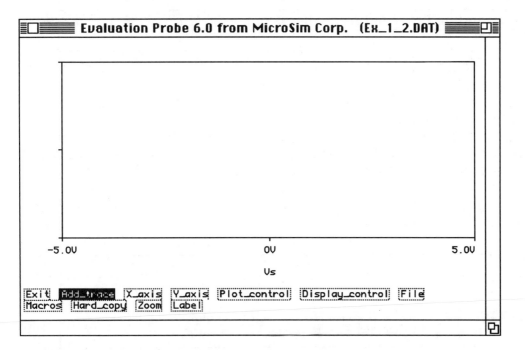

FIGURE 1.20 Probe at Startup After Selection of EX_1_2.DAT

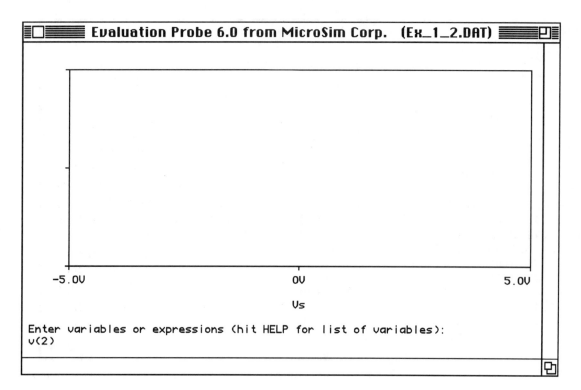

FIGURE 1.21 After Typing v(2).

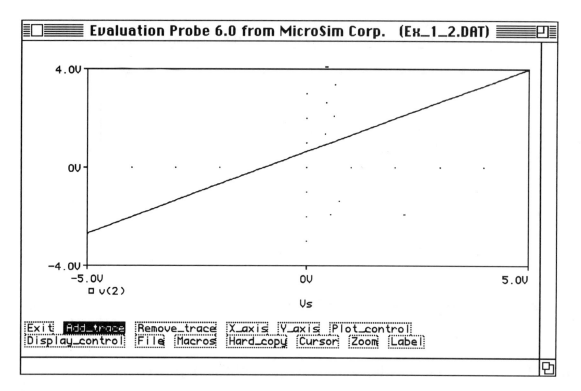

FIGURE 1.22 Plot of $V_2$ vs. $V_S$

With either an IBM PC or a Macintosh you now see the Analog/Digital menu window shown in Fig. 1.20.

Since $V_S$ varies from -5 V to 5 V, Probe already plans to plot any dependent variable as a function of the source voltage $V_S$. Because **Add_trace** is highlighted, you can select this item by typing <Return>. Alternately, you can type the character A or click on the **Add_trace** item with the mouse. After selecting **Add_trace**, Fig. 1.21 shows the window after you type V(2). You also can enter V(2) with an IBM PC by typing the function key <F4> and selecting V(2) with the arrow keys or mouse from the list of variables that appears at the top of the screen. With a Macintosh you can use the <Help> key or select **Help** on the Traces menu in the menubar and select V(2) with the arrow keys or mouse, or you can select **Analog Node Voltages** from the Traces menu on the menubar and select V(2). After V(2) appears on your screen, type <Return> to have a plot of $V_2$ as a function of $V_S$ appear in the Probe plot window, as shown in Fig. 1.22. Try adding a plot of node-voltage difference $V_1$ - $V_2$ as a function of $V_S$ to your plot, again using **Add_trace**. Type either V(1,2) or V(1)-V(2) to tell Probe that you want $V_1$ - $V_2$ to appear in the plot. The plot shows a linear dependence of $V_2$ on the voltage source $V_S$. Also, from the output list you can see that when $V_S$ equals zero $V_2$ = -($V_1$ - $V_2$) = 0.6667 V, which is the value due to current source $I_S$ alone.

Chapter 2, Section 6 describes operation of the other menu items on the Analog/Digital menu. In the meantime, you can explore using the other Analog/Digital menu items. To select one of the menu items use either of the following methods:

- Type the capital letter in the item, e.g., type R to select **Remove_trace**.

- Use the right <→> or left <←> arrow keys to change the item that appears in highlight, then type <Return> or <Enter>.

- Click on the item with the mouse.

Typing the <Escape> key or clicking the mouse on **Exit** is equivalent to typing character E. Macintosh users can select Analog/Digital menu items using the mouse to select the item from the Probe menu on the menubar instead of clicking on the item at the bottom of the screen.

# Chapter 2 How to Use PSpice and Probe

This chapter describes how to use PSpice and Probe to analyze and simulate electrical circuits. Some of the commands are specific to PSpice, but most of the format descriptions apply to all versions of SPICE. PSpice exists for many different computer systems, including the IBM PC and Macintosh. The evaluation version of PSpice restricts the size of the circuit but has all of the features of the commercial version. The evaluation version of PSpice is in the public domain, so you can copy and distribute it.

The first part of this chapter describes the syntax of PSpice circuit netlists. Subsequent sections describe DC, AC, and transient circuit analysis commands. The last section on PSpice describes analog behavioral modeling, use of subcircuits and libraries, creation of functions, Monte Carlo statistical analysis, Fourier analysis, mutual inductance, and operating point control. The final section in this chapter explains how to use Probe effectively. In this section you will learn about MAC and GF files, which let you create macros and goal function procedures. Axis-, plot-, and cursor-control features of Probe let you prepare Probe plots for effective display. With Probe's performance analysis and goal functions you can measure circuit specifications and plot these as functions of a circuit parameter. Probe's Fourier command changes the time-axis display of transient variables to a Fourier transform spectrum. The histogram feature plots statistical circuit behavior in a bar-graph mode.

## 2.1 Introduction

You need to describe your circuit to PSpice and specify the analyses that you wish. To do this you create and save a netlist as an ASCII text file, using any suitable word-processor program. You name this netlist using a name of your choice, usually with a CIR suffix. The PC version of PSpice requires a CIR suffix with the file name. The Macintosh version of PSpice does not require a CIR suffix, but using this suffix helps to identify the file as a PSpice netlist. PSpice creates an output file that contains the circuit simulation results. This output file has the same name as your netlist file, but the suffix is OUT. When your netlist includes a PROBE statement, PSpice creates a second output file for use by the Probe application. This file has the same name as your netlist, but the suffix is DAT.

The CIR netlist includes some or all of the following items:

- A first title line
- Option statements
- Control statements
- Comment lines
- Lines that name circuit elements, nodes, and values
- Analysis specifications
- Output specifications
- Model statements
- Subcircuit statements
- Parameter statements

- Function statements
- A closing END line

The first line of every netlist has to be a title line. If you do not write a title line, PSpice thinks that the next line is your title. This error usually leads to results that you do not intend. This first line can be blank, but documentation of the netlist with a meaningful title is very useful. If you wish, you can begin this title line with an asterisk (*) in the first column, which indicates a comment. However, the title line does not need to begin with an asterisk.

SPICE manuals tell you that every netlist terminates with an END statement. In fact, PSpice programs run without an END statement. However, good practice dictates that you include the END statement. A useful feature of the END statement is that another netlist can follow a previous netlist that terminates with an END statement. PSpice analyzes all netlists in the file as separate jobs without further intervention by you. The END statement is simply

.END

The decimal point before END shows that this statement is a command-line statement. If you have any blank lines after an END statement, the output file gives normal results and also tells you that you have a circuit with no elements. Although this message may be confusing, it simply confirms the existence of these blank lines.

Except for the title line at the beginning, the closing END line at the end, subcircuit descriptions that cannot contain any main netlist statements, and function statements that have to precede any reference to them, SPICE statements can be in any order.

SPICE does not differentiate between lower- and uppercase alphabetic characters. Any statement following an asterisk (*) as the first character in a line is a comment statement that SPICE ignores. In PSpice, all characters following a semicolon also are comments. A plus sign (+) at the start of a line means that the previous line continues on the current line.

The nodes of a circuit usually have number identifiers and these identifiers need not be sequential. The ground node always is node 0. In PSpice, nodes also can have alphanumeric names, using up to 131 sequential characters. PSpice nodes may be global. Global nodes are accessible within subcircuit descriptions without appearing in the subcircuit statement. A global node must have the prefix $G_, e.g. $G_V+ could name a positive supply voltage terminal.

Before proceeding with a simulation, SPICE checks the circuit. There has to be a DC path from every node to the ground node. A dependent or independent current source, capacitance, or the input terminals of a voltage-controlled dependent source do not provide a path to ground. Also, a circuit may not contain a loop of independent or dependent voltage sources or inductances.

SPICE also checks to make sure that every node connects two or more circuit elements. As an exception, PSpice allows you to connect an independent or a dependent voltage source from a node in your circuit to another node that has no other attached circuit elements. PSpice adds a resistance of value 1/GMIN in parallel with the voltage source. By default, parameter GMIN equals $10^{-12}$ S, and 1/GMIN is the largest resistance that

| F | femto | $10^{-15}$ |
|---|-------|-----------|
| P | pico | $10^{-12}$ |
| N | nano | $10^{-9}$ |
| U | micro | $10^{-6}$ |
| M | milli | $10^{-3}$ |
| K | kilo | $10^{3}$ |
| MEG | Mega | $10^{6}$ |
| G | Giga | $10^{9}$ |
| T | Tera | $10^{12}$ |

Table 2.1 SPICE Scaling Factors

PSpice uses. PSpice does allow you to connect both terminals of a two-terminal element to the same node, creating a trivial second circuit.

You can write rational numbers using floating point notation, exponential (E) notation, or scaling factors. Table 2.1 shows SPICE scaling characters. You can append these characters to a number, without any intervening blank space. For example 0.0012, 1.2M, and 1.2E-3 are equivalent representations of the numeral $1.2 \times 10^{-3}$. Since lower- and uppercase characters are the same to SPICE, the characters m and M both mean milli.[1] For this reason, the Mega scaling factor is the three-character combination MEG or Meg. You can write units following an element's numerical value, but do not leave a blank space between the number and the unit. Using Farad as a unit name without a preceding scaling factor may be a serious error! However, using Ampere or A works, because atto ($10^{-18}$) does not exist in SPICE.

| Option | Meaning |
|--------|---------|
| ACCT | Calls for accounting information to appear |
| EXPAND | Shows subcircuit expansions |
| LIBRARY | Lists lines from libraries |
| LIST | Lists circuit elements |
| NOBIAS | Suppresses node-voltage bias values |
| NODE | Lists the node table |
| NOECHO | Suppresses listing of the input file |
| NOMOD | Suppresses listing of model parameters and temperature changes |
| NOOUTMSG | Suppresses error messages in output file |
| NOPAGE | Suppresses paging and banner production |
| NOPRBMSG | Suppresses error messages in DAT file |
| NOREUSE | Suppresses reuse of bias point |
| OPTS | Lists values for options |

Table 2.2 Flag Options

---

[1] Probe uses the upper case character M for mega and lower case m for milli.

| Option | Meaning | Unit | Default |
|---|---|---|---|
| ABSTOL | Accuracy of currents | A | 1 p |
| CHGTOL | Accuracy of charge | C | 10 f |
| CPTIME | CPU time for run | s | 1 M |
| DISTRIBUTION | Default Monte Carlo distribution | | UNIFORM |
| GMIN | Minimum branch conductance | S | 1 f |
| ITL1 | DC and bias blind iteration limit | | 40 |
| ITL2 | DC and bias educated guess iteration limit | | 20 |
| ITL4 | Transient iteration point limit | | 10 |
| ITL5 † | Total transient iteration limit (ITL5=0 means ∞) | | ∞ |
| LIMPTS | Maximum print or plot points (LIMPTS=0 means ∞) | | ∞ |
| NUMDGT | Significant digits (< 8) | | 4 |
| PIVREL † | Relative pivot magnitude | | 1 m |
| PIVTOL † | Absolute pivot magnitude | | 100 a |
| RELTOL | Relative V and I accuracy | | 1 m |
| TNOM | Nominal temperature | °C | 27 |
| VNTOL | Accuracy of voltage values | V | 1μ |

Table 2.3 Value Options († Change these at your own risk)

The SPICE operation environment can be set using the OPTIONS (or OPT) statement. The syntax[2] of this statement is

.OPTIONS [<Option_Name>] ... [<Option_Name>=<Value>] ...

Some options behave as flags, others set values. Table 2.2 lists flag options that you can set, and Table 2.3 shows options that you set to a value or a name. You can list the options in any order. Options that set values use the equality (=) sign between the option name and <Value>. For readability, you can use additional spaces before and after the equality sign.

## 2.2 DC Analysis with SPICE

This section shows how to simulate circuits that have DC sources. Some of the topics apply to AC and transient simulations as well as to DC simulations. In particular, with some modification, a device specification applies to AC or transient simulations. Also, the STEP, TEMP, and PARAM commands affect AC and transient simulations. For now, only the DC affect of these device and command statements are of interest.

---

[2] PSpice netlist syntax descriptions in this manual use the following notation. Replace character sequences having <> delimiters and the <> delimiters with appropriate character sequences that you type. Square brackets [] indicate optional items. Do not type these braces, except where their presence indicates alphanumeric nodes in output statements. Ellipses "..." indicate that you can repeat previous items.

## 2.2.1 Simple Device Specification

This section describes the simplest forms of two-terminal device specifications that you can use in a SPICE netlist. These simple element forms include resistance, inductance, capacitance, and independent and dependent voltage and current sources.

### *Resistance*

To specify a resistance that connects between <Node_1> and <Node_2> and has the value <Value> write a statement having the form

      R<Name> <Node_1> <Node_2> <Value>

Replace <Name> by almost any alphanumeric name that you wish, using no more than 131 characters. However, keeping device names to eight characters improves performance. Usually you will use a name that relates to the literal name of the element in the circuit. For example, resistance $R_1$ in a circuit becomes R1 in the SPICE netlist. The order of the entries for <Node_1> and <Node_2> determines the polarity for current and voltage. The positive direction for current is from <Node_1> to <Node_2>. SPICE uses the passive convention for voltages and currents, so SPICE measures positive voltage drop also from <Node_1> to <Node_2>. The entry <Value> is the numerical value of the resistance. For example, if a 2000-$\Omega$ resistance $R_5$ connects between nodes 3 and 6, you can type

      R5 3 6 2000

to describe resistance $R_5$ in the netlist. The following lines are equivalent to the line above:

      R5 3 6 2K

      r5 3 6 2kOhm

Typing

      R5 6 3 0.002Meg

reverses the definition of current direction and voltage drop for resistance $R_5$.

### *Inductance*

To represent an inductance element in a netlist use

      L<Name> <Node_1> <Node_2> <Value> [IC=<IC_Value>]

Again, select <Name> to relate to the circuit diagram. The sequence <Node_1>, <Node_2> sets both the positive current direction and voltage drop polarity in the inductance. Enter the inductance value in place of <Value>. Replace <IC_Value> by the initial inductance current. If you omit IC=<IC_Value> the initial current defaults to zero. The initial current is positive when its direction is from the first node number toward the second.

## Capacitance

To represent a capacitance element in a netlist use

C<Name> <Node_1> <Node_2> <Value> [IC=<IC_Value>]

Again, select <Name> to relate to the circuit diagram. The sequence <Node_1>, <Node_2> sets the direction of positive voltage drop and current polarity in the capacitance. Enter the capacitance value in place of <Value>. Replace <IC_Value> by the initial capacitance voltage. If you omit IC=<IC_Value> the initial voltage defaults to zero. The initial voltage drop is positive when the first numbered node is at a higher potential than the second.

## Independent Voltage Source

An independent DC voltage-source statement has the form

V<Name> <Node_1> <Node_2> [DC] [<Value>]

Choose <Name> to relate to the name appearing in your circuit diagram. The <Node_1>, <Node_2> order usually follows the + and - labels in your circuit diagram. The square brackets about DC and <Value> imply that these entries are optional. Entering DC before <Value> is optional. A source defaults to DC if you omit these characters. Omitting both DC and <Value> sets the source to zero. To reverse the order of the nodes, change the sign of <Value>. Either

VS 5 4 DC 3.5

or

Vs 4 5 -3.5

describes a 3.5-V DC source V$_S$ having a positive voltage at node 5 with respect to node 4.

## Independent Current Source

An independent current source statement has the form

I<Name> <Node_1> <Node_2> [DC] [<Value>]

Replace <Name> by characters of your choice. The positive direction of the current source is from <Node_1> to <Node_2>. DC and <Value> are both optional. The current source value defaults to zero if you omit <Value>.

## Controlled Voltage and Current Sources

Voltage-controlled voltage source (VCVS), current-controlled current source (CCCS), voltage-controlled current source (VCCS), and current-controlled voltage sources (CCVS) statements have the form

E<Name> <Node_1> <Node_2> <Node_3> <Node_4> <Value>

F<Name> <Node_1> <Node_2> V<Cont> <Value>

G<Name> <Node_1> <Node_2> <Node_3> <Node_4> <Value>

H<Name> <Node_1> <Node_2> V<Cont> <Value>

<Name> identifies the element in the circuit. <Node_1> and <Node_2> give the nodes of the controlled source in the circuit and show the polarity of the current or voltage. The voltage-controlled sources (E and G) indicate the controlling-voltage node-pair and polarity with <Node_3> and <Node_4>. The current in the V<Cont> independent voltage source is the controlling current for the current-controlled sources (F and H). If necessary, add a null-valued DC source to the circuit to measure the controlling current, but this source need not be zero. <Value> is the scale factor that relates the value of the dependent source to the controlling variable. Unfortunately, you cannot use parameters (see Sec. 2.2.10) as <Value> in this form of dependent source.

### 2.2.2 General R, L, and C Models

The simple R, L, and C element representations above often suffice. However, these are special cases of the general SPICE element representations. These allow for temperature dependence in R, L, and C devices. With the general models, inductance L can have current nonlinearity, and capacitance C can have voltage nonlinearity. To model nonlinear resistance, use VCVS- and VCCS-type dependent sources (See Section 2.5.1). Use these general device forms, rather than the simpler ones in Sec. 2.2.1, if you wish to sweep an element model parameter in an output statement.

### *Resistance*

The general resistance statement in SPICE has the form

    R<name> <Node_1> <Node_2> [<Model_Name>] <Value>
    +      [TC=<TC1>, [<TC2>]]

<Model_Name> is an alphanumeric name of your choice that identifies a MODEL statement somewhere else in the netlist by appearing in that statement as well. Giving <Model_Name> without specifying TCE in a MODEL statement, causes the resistance value to be

$$<Value> * R * [1 + TC1 * (T - TNOM) + TC2 * (T - TNOM)^2]$$

With no model specification and in-line temperature coefficients, SPICE uses unity as the parameter value R. The MODEL statement gives values for any or all of parameters R, TC1, TC2, and TCE. Table 2.4 lists the default values of the resistance model parameters. The MODEL statement values of TC1 or TC2 override the in-line values. If the MODEL statement gives TCE, then the resistance value is

| Parameter | | Units | Default |
|---|---|---|---|
| R | Resistance multiplier | | 1 |
| TC1 | Linear temperature coefficient | $°C^{-1}$ | 0 |
| TC2 | Quadratic temperature coefficient | $°C^{-2}$ | 0 |
| TCE | Exponential temperature coefficient | $%/°C^{-1}$ | 0 |

Table 2.4 Resistance Model Parameters

$$<Value> * R * 1.01^{[TCE * (T - TNOM)]}$$

<Value> normally assumes positive values, can be negative, but never becomes zero. Set TNOM using the OPTIONS (use OPT to abbreviate) statement

.OPT TNOM = <Value>

The nominal temperature TNOM defaults to 27 °C.

## *Capacitance*

The general capacitance statement in SPICE has the form

```
C<name> <Node_1> <Node_2> [<Model_Name>] <Value>
+     [IC=<IC_Value>]
```

Again, <Model_Name> associates the capacitance statement with its model. If you omit <Model_Name> the capacitance value is <Value>. Giving <Model_Name> causes the capacitance value to be

$$<Value> * C * (1 + VC1 * V + VC2 * V^2)$$
$$* [1 + TC1 * (T - TNOM) + TC2 * (T - TNOM)^2]$$

The MODEL statement can give a value for any of C, VC1, VC2, TC1, or TC2. Table 2.5 lists the default values of C, VC1, VC2, TC1, and TC2. <Value> normally assumes positive values, can be negative, but never becomes zero.

## *Inductance*

The general inductance statement in SPICE has the form

```
L<name> <Node_1> <Node_2> [<Model_Name>] <Value>
+     [IC=<IC_Value>]
```

<Model_Name> relates the inductance statement to an inductance model statement. If you omit <Model_Name> the capacitance value is <Value>. Giving <Model_Name> causes the inductance value to be

$$<Value> * L * (1 + IL1 * I + IL2 * I^2)$$
$$* [1 + TC1 * (T - TNOM) + TC2 * (T - TNOM)^2]$$

The MODEL statement can give a value for any of L, IL1, IL2, TC1, or TC2. Table 2.6 lists the default values of L, IL1, IL2, TC1, and TC2. <Value> normally assumes positive values, can be negative, but never becomes zero.

| Parameter | | Units | Default |
|---|---|---|---|
| C | Capacitance multiplier | | 1 |
| VC1 | Linear voltage coefficient | $V^{-1}$ | 0 |
| VC2 | Quadratic voltage coefficient | $V^{-2}$ | 0 |
| TC1 | Linear temperature coefficient | $C^{-1}$ | 0 |
| TC2 | Quadratic temperature coefficient | $C^{-2}$ | 0 |

Table 2.5 Capacitance Model Parameters

| Parameter | | Units | Default |
|---|---|---|---|
| L | Inductance multiplier | | 1 |
| IL1 | Linear voltage coefficient | $A^{-1}$ | 0 |
| IL2 | Quadratic voltage coefficient | $A^{-2}$ | 0 |
| TC1 | Linear temperature coefficient | $C^{-1}$ | 0 |
| TC2 | Quadratic temperature coefficient | $C^{-2}$ | 0 |

Table 2.6 Inductance Model Parameters

### 2.2.3 The MODEL Statement

The MODEL statement is necessary when a device refers to a model name, as in the case of the general element forms above. The format of the MODEL statement is

```
.MODEL <Model_Name> [AKO: <Ref_Model_Name>] <Type_Name>
+     [(([<Param_1>=<Value_1> [<Tolerance_Spec>]] ...
+     [T_MEASURED=<T_Measured>]
+     [T_ABS=<T_Abs>]/T_REL_GLOBAL=<T_Rel_Global>/
+     T_REL_LOCAL=<T_Rel_Local>])]
```

<Model_Name> is the name that a device statement (R, L, C, etc.) uses to refer to a specific model. <Model_Name> starts with an alphabetic character. Using the letter preceding <Name> in the device statement as the starting letter in the <Model_Name> is common, since this practice makes the netlist more readable. For example

```
R1 1 2 Rsample 1k
...

...
.MODEL Rsample RES (R=2.2, TC1=12.5m)
```

Using the AKO (A Kind Of) option allows you to refer to another model of the same device type. Replace <Ref_Model_Name> with the name of another model that serves as a base for the model in this statement. Parameters in the model having <Ref_Model_Name> apply, unless you explicitly override them in the current model. Replace <Type_Name> with the appropriate type name for the device from the first column of Table 2.7. Following <Type_Name>, place any parameter specifications. You may specify none, any, or all parameters in any order. Parameters that you do not explicitly specify assume their default values (See Tables 2.4, 2.5, and 2.6 for R, L, and C default parameter values). A model parameter can be swept in a DC or STEP statement.

<Tolerance_Spec> may include

```
[DEV[<L_&_D>] <value>[%]] [LOT[<L_&_D>] <value>[%]]
```

DEV applies to a specific device, LOT to a collection of devices. The percent sign (%) indicates that the value is a percentage, otherwise the value has the parameter's normal units. The optional <L_&_D> specification can include either or both of

```
[/<Lot_No>][/<Distribution_Name>]
```

| Type Name | Character | Device |
|---|---|---|
| RES | R | Resistance |
| CAP | C | Capacitance |
| IND | L | Inductance |
| D | D | Diode |
| NPN | Q | npn bipolar junction transistor (BJT) |
| PNP (LPNP) | Q | (lateral) pnp bipolar junction transistor |
| NJF | J | n-channel junction field-effect transistor (JFET) |
| PJF | J | p-channel junction field-effect transistor |
| NMOS | M | n-channel metal-oxide semiconductor field-effect transistor (MOSFET) |
| PMOS | M | p-channel metal-oxide semiconductor field-effect transistor |
| GASFET | B | n-channel GaAs field-effect transistor (MESFET) |
| CORE | K | nonlinear magnetic core (transformer) |
| VSWITCH | S | voltage-controlled switch |
| ISWITCH | W | current-controlled switch |

Table 2.7 <Type_Name> and Character Symbols for MODEL Statement

Replace <Lot_No> with an integer from 0 to 9. Each integer specifies a different random number generator. Using this option allows correlation between parameters in the same model and between models. DEV and LOT variations are independent and use one of ten different random numbers for DEV and for LOT. If you do not use <Lot_No> then each parameter uses a different random number. <Distribution_Name> supplies the name of the statistical distribution to use when PSpice calculates the actual parameter from the parameter value in the MODEL command. You can let <Distribution_Name> be UNIFORM, GAUSS, or a user name. (See Sec. 2.5.5 for details.)

For the general R, L, and C models, you can set T_MEASURED = <Value> in the list of parameters in a MODEL statement to indicate that the model parameters apply at the T_MEASURED temperature in °C. If you enter T_ABS=<Value> in the parameter list, elements using this model have this temperature in °C instead of the default temperature TNOM. Specifying T_REL_GLOBAL=<Value> as a MODEL statement parameter indicates that devices based on this model have temperatures equal to TNOM + T_REL_GLOBAL. Finally, setting T_REL_LOCAL = <Value> in a MODEL statement indicates that devices having this model operate at T_ABS(AKO Model) + T_REL_LOCAL. You can use only one of T_ABS, T_REL_GLOBAL, or T_REL_LOCAL.

### 2.2.4 The OP Statement

Analysis defaults to OP when no other analysis statements appear in the netlist. Using OP makes PSpice provide operating-point information in your output file about node voltages, current, and power for each independent voltage source and small-signal linearized parameters for all nonlinear controlled sources or semiconductor devices. To use this command, just type the line

    .OP

somewhere in your netlist.

### 2.2.5 The DC Statement

The DC statement has the forms

```
.DC [LIN] <Sweep_Var> <Start_Value> <End_Value> <Inc>
+       [Nested_Sweep]

.DC OCT  <Sweep_Var> <Start_Value> <End_Value> <Points>
+       [Nested_Sweep]

.DC DEC  <Sweep_Var> <Start_Value> <End_Value> <Points>
+       [Nested_Sweep]

.DC  <Sweep_Var> LIST <Value_1> <Value_2> ...
+       [Nested_Sweep]
```

<Sweep_Var> can be a source name, model parameter [e.g., RES R<model>(R), where R<model> is the model name], temperature (TEMP), or a global parameter (e.g., PARAM <Param_Name> – see Sec. 2.2.10 ). <Start_Value> and <End_Value> define the starting and ending values of <Sweep_Var>. For a linear sweep <Inc> is the increment of the sweep variable. PSpice allows a linear sweep from larger to smaller values by setting <End_Value> smaller than <Start_Value>. However, for either a positive- or a negative-direction sweep, <Inc> must be positive. For the logarithmic sweep types (OCT and DEC), where the sweep variable changes by equal-valued logarithmic increments, <Points> is the number of points per octave or decade. Each successive value exceeds the previous value by a factor equal to ten (DEC) or two (OCT) to the power 1/<Points>. SPICE stops the sweep at the last value remaining less than or equal to <End_Value>. For example, if your netlist includes the DC statement

```
.DC   DEC  Vs   0.8   5.0   5
```

SPICE simulates the circuit for values of voltage source $V_s$ equal to 0.8, 1.268, 2.010, and 3.185 V, because 10^(1/5) equals 1.58489 and 5.050, the next value in the sequence, exceeds 5.0. The values that you enter for <Start_Value> and <End_Value> have to be positive and <End_Value> has to exceed <Start_Value>. The optional nested sweep can be any of the four types of sweep (LIN, OCT, DEC, or LIST) involving a second sweep variable. With a nested sweep, the first variable is the inner loop and cycles more rapidly. You cannot nest beyond a second variable, and SPICE allows only one DC statement in a netlist.

If you want to obtain more information about a circuit solution than OP gives, use a PRINT DC statement with a single-valued sweep. In the single-valued sweep you select any one of your independent sources and refer to it in a sweep statement that reproduces that source's value for one step. For example, suppose that you have a voltage source $V_{S1}$ = 5 V connecting between node 1 and node 0 and a current source $I_{S1}$ = 2 A connecting from node 0 to node 5 and a resistance $R_3$ = 2.2 kΩ connecting between node 3 and node 4, whose current is to appear in the output file. The following netlist fragment shows how to measure the current $I_{R3}$ in resistance $R_3$.

```
...
VS1   1    0    5
IS1   0    5    2
...
.DC   LIN   VS1  5    5    1
.PRINT DC   I(R3)
...
```

In this example, the DC statement causes $V_{S1}$, chosen arbitrarily over $I_{S1}$, to have a one-step sweep consisting of the single 5-V value. The increment in this example is 1 V, but can be any other positive number. Without the DC statement SPICE ignores the PRINT DC statement and produces only the operating point values by default. See Sec. 2.2.9 for further details about the PRINT statement and how to specify output variables.

### 2.2.6 The STEP Statement

The STEP command applies only with PSpice, not with SPICE. With DC analysis, the STEP statement is an alternative to a nested DC specification. However, the resulting output has better documentation of the additional variable steps. Also, the STEP statement has a profound significance when used with either AC or TRAN analysis commands (See Secs. 2.3 or 2.4), as these also repeat for each step of the variable in the STEP command. The STEP command has the same sweep forms as the DC command (replace DC with STEP). PSpice allows only one STEP statement in a netlist.

### 2.2.7 The TEMP Statement

With the TEMP command you can set the temperature to assume one or more values. With more than one value, all other commands repeat for each temperature. The form of this command is

.TEMP <Temp_Value_1> ...

### 2.2.8 The TF Statement

The TF statement has the format

.TF <Output_Var> <Input_Src>

and causes SPICE to determine the small-signal transfer function relating the output variable <Output_Var> to the input source <Input_Src>. Also, SPICE calculates the input and output resistances. When the output variable is a current, you have to specify this current as the current in an independent voltage source. This command is most useful for measurement of amplifier characteristics, but remember that it applies only to DC analysis.

### 2.2.9 The PRINT DC, PLOT DC, and PROBE Statements

SPICE's PRINT DC, PLOT DC, or PSpice PROBE commands generate specific types of DC output. The PRINT DC statement produces DC output values in tabular form, the PLOT DC command creates line plots of DC output values, and the PROBE command creates a data file that the Probe application uses to create output plots. PSpice gives this data file the same name as your netlist file but replaces the CIR suffix with DAT. These commands have the form

.PRINT DC <Output_1> ...

.PLOT DC <Output_1> ...

.PROBE [<Output_1> ...]

Each output specification, e.g., <Output_1>, identifies an output voltage or current that you want to appear in the output file or to be available in Probe. An output voltage specification names a node voltage, e.g., V(3) for node 3, or a node voltage difference, e.g., V(1,3) for the voltage difference between node 1 and node 3. PSpice, but not Probe, also lets you name a two-terminal element to refer to the voltage across that element. For example, V(R3) measures the voltage across resistance $R_3$. To obtain a current value as an output, name the current in a voltage source in the output specification. For example, the current in a voltage source named VEXAMPLE is I(VEXAMPLE). If necessary, insert a null-valued voltage source into your circuit as an ammeter to measure a current. PSpice and Probe allow you to refer to currents in two-terminal elements directly; e.g., I(R1) is the current in resistance $R_1$ having the direction from the first to the second node of $R_1$. If a node has an alphanumeric name, you must place square brackets [ ] around the node name in an output specification. For example, to specify the voltage at node OUT use V([OUT]) in the PRINT, PLOT, or PROBE statement. If you do not include the square brackets, PSpice thinks OUT is a device and sends you a diagnostic error message.

There is no limit to the number of variables in a PRINT statement. The output file breaks the data at an appropriate width. You can force different sets of data to appear in groups by having more than one PRINT command. The PLOT command can refer to no more than eight variables. However, a netlist can have more than one PLOT command. The PROBE command has no limit to the number of variables that can be given. If the PROBE statement has no specification of output variables, then all circuit variables are available from within Probe. The reason for having output variables in a PROBE statement is to limit the number of variables to those of interest in order to limit the size of the DAT file that PSpice produces. Of course, there is no point in having more than one PROBE statement in a netlist. Unlike PRINT DC and PLOT DC, DC does not appear in the PROBE command. There is no point in having a PRINT DC or PLOT DC command without also having a DC command in the netlist.

### 2.2.10 The PARAM Statement

In PSpice you can define parameters using the PARAM statement

.PARAM <Var_Name>=<Value> ...

.PARAM <Var_Name>={<Expression>} ...

Subsequently you can use <Var_Name> within curly braces {} instead of a numerical value. You also can include explicit calculations within curly braces. In addition to the arithmetic operations +, -, *, and /, and the use of parentheses, the functions in Table 2.8 are available for computations within curly braces.

| Function | Expression | Comment |
|---|---|---|
| abs(x) | \|x\| | |
| exp(x) | exp(x) | |
| sin(x) | sin(x) | x in radians |
| cos(x) | cos(x) | x in radians |
| tan(x) | tan(x) | x in radians |
| atan(x)=arctan(x) | $\tan^{-1}(x)$ | result in radians |
| log(x) | ln(x) | log base e |
| log10(x) | log(x) | log base 10 |
| pwr(x,y) | $\|x\|^y$ | |
| pwrs(x,y) | $\|x\|^y$ (x > 0), $-\|x\|^y$ (x < 0) | |
| sqrt(x) | $\sqrt{x}$ | |
| table(x,$x_1$,$y_1$,...,$x_n$,$y_n$) | | $y_1$ for x<$x_1$, $y_n$ for x>$x_n$ interpolates for $x_1$<x<$x_2$ |
| limit(x, min, max) | | min for x<min, max for x>max, otherwise x |

Table 2.8 PSpice Expressions

You can use parameters, calculations, and functions instead of values in most device statements, but not to represent node numbers, values in analysis statements (e.g., DC), or values of E, F, G, or H sources. The analog behavioral modeling possible with E- and G-type sources (see Sec. 2.5.1) overcomes this unfortunate restriction on the use of parameters in E, F, G, or H sources.

## Example 2.1 DC Bridge Simulation

This example uses the DC bridge circuit in Fig. 2.1 to illustrate how to use both a resistance model and the PARAM statement to sweep and step two variables. In this circuit, resistance $R_M$ is a meter resistance and resistance $R_X$ is an unknown resistance that we wish to measure. Bridge resistance $R_B = R_2 = R_3 = R_4$ has a 10-k$\Omega$ value. The task in this example is to use Probe to display the meter current $I_{RM}$ as a function of the base-10 logarithm of $R_X$, with meter resistance values of 100 $\Omega$, 1 k$\Omega$, and 10 k$\Omega$. In addition, we will plot an approximation of the solution of $I_{RM}$ for the case where $R_M$ is much smaller than 10 k$\Omega$.

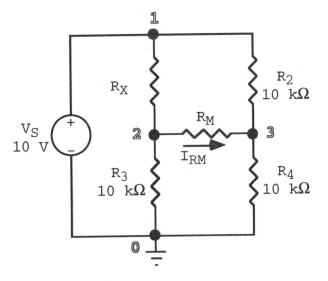

FIGURE 2.1 DC Bridge Circuit

## Solution

Figure 2.2 shows a PSpice netlist to solve this problem. This solution uses resistance model Rmeas to represent resistance $R_X$. Then, the DC statement sweeps the R value of this model in 10 steps per decade from 0.1 to 10. This sweep leads to resistance values of $R_X$ starting at 1 k$\Omega$ and ending at 100 k$\Omega$ with ten points per decade. Choosing <Value> equal to 10 k in the Rx statement makes parameter R in the model become the fraction of the resistance value of $R_X$ relative to the bridge resistance of 10 k$\Omega$. The STEP statement using the LIST form causes PSpice to repeat the DC analysis three times: once with $R_M$ equal to 100 $\Omega$, a second time with $R_M$ equal to 1 k$\Omega$, and finally with $R_M$ equal to 10 k$\Omega$. The PROBE statement makes available all circuit variables for plotting when Probe runs. The Probe plot of $I_{RM}$ vs. log(R) results from selection of **All_Dc_sweep** after opening the EX_2_1.DAT file and choosing **Add_trace** and entering I(Rm). You can either type i(rm) or select I(Rm) from the screen of variables that appears when you type the <Help> key or select **Help** from the **Traces** menu. Use the cursor or arrow keys

```
Example 2.1 - DC Bridge Simulation
.PARAM Rval = 100
.MODEL Rmeas RES   ; Use default resistance parameters
Vs 1 0 10V
Rx 1 2 Rmeas 10k
R2 1 3 10k
R3 2 0 10k
R4 3 0 10k
Rm 2 3 {Rval}
.DC DEC RES Rmeas(R) 0.1 10.0 10
.STEP PARAM Rval LIST 100 1k 10k
.PRINT DC I(Rx) V(2,3)
.PROBE
.END
```

FIGURE 2.2 DC Bridge Simulation Netlist

to highlight the selection, type <Enter> or <Return> to insert the selection of I(Rm) into the **Add_trace** entry space, and type <Enter> or <Return> once again to complete the process. Change the R axis to a log plot by selecting **X_axis** from the Analog/Digital menu and **Log** from the **X_axis** menu.

When $R_M$ is small compared to bridge resistance $R_B = R_2 = R_3 = R_4 = 10\ k\Omega$, the current $I_{RM}$ is

$$I_{RM} \approx \frac{1 - (R_x / R_B)}{1 + 3(R_x / R_B)} \times \frac{V_s}{R_B}$$

To plot this approximate solution using Probe, select **Add_trace** and type the expression shown in Fig. 2.3. Since the netlist in Fig. 2.2 has <Value> = 10 kΩ, parameter R in the resistance model is the ratio of $R_X$ to $R_B$. You can refer to parameter names in Probe expressions to create plots. The solution for the case where $R_M$ equals 100 Ω and the approximate solution are nearly the same. However, the other two solutions differ quite a bit, since in these cases meter resistance $R_M$ equals and greatly exceeds the 10-kΩ resistance value.

As an alternative to using the STEP command to sweep the meter resistance, try using a nested sweep specification. Remove the Step statement and change the DC statement to

.DC DEC RES Rmeas(R) 0.1 10. 10 PARAM Rval LIST 100 1k 10k

Now, run PSpice with this new version of the netlist. Compare the output files and the way Probe behaves for each case. You may conclude that this method gives the same results, but the documentation of the three $R_M$ cases is not as good using the nested sweep. Also, you can explore doing both sweeps with parameter values or doing both with resistance models.

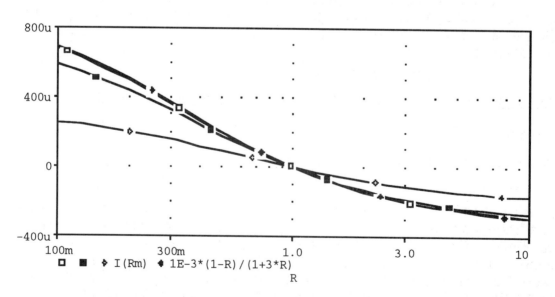

FIGURE 2.3 Current in $R_M$ vs. $\log(R_X)$

## 2.3 AC Analysis with SPICE

The following sections explain how to use SPICE to do AC analysis. Before doing an AC analysis, SPICE finds the DC operating-point values for the circuit variables. Next, SPICE linearizes the circuit about this operating point and does an AC phasor analysis using the circuit's operating-point linearization. For a linear circuit this two-step process is equivalent to setting all DC sources to zero and then doing an AC analysis. However, when the circuit contains nonlinear elements the difference between these two processes is significant.

### 2.3.1 AC Independent Source

To have an AC component in an independent voltage or current source, append the specification

　　　AC <Magnitude> [<Phase>]

following a DC specification, if any. The AC part of the source applies only to AC analysis. Any DC specification determines the DC operating point for an AC analysis, but this fact has no significance for linear circuits.

<Magnitude> is either the maximum amplitude or the RMS magnitude of the source. If you use maximum amplitude values for voltage or current sources, then all responses are maximum amplitudes. If you enter RMS amplitude values for voltage or current sources, then all responses are RMS amplitudes. If you specify RMS values for some sources and maximum values for other sources in the same netlist your results will be incorrect.

<Phase> is the phase shift of the source in degrees. If you omit <Phase>, then the phase of the source defaults to $0°$. SPICE does not know whether you intend a sinusoidal source in the time-domain circuit to be a sine or cosine function. If your sources are all cosine functions, then the responses also are all cosine functions. Obviously, in any one netlist you must represent all sources in the same way, using only all-cosine or all-sine function representation.

As an example of an independent voltage-source specification, if an independent voltage source $v_{S1}$ connects between node 1 and node 0 and has a DC value of 5 V and an AC value of $7.5\angle-58°$ V, the SPICE description of this element is

　　　VS1 1 0 DC 5 AC 7.5 -58

In this statement you can omit DC without affecting the SPICE solution.

### 2.3.2 The AC Statement

The syntax of a statement for AC analysis is

　　　.AC [LIN] [OCT] [DEC] <Points> <Start_Freq> <End_Freq>

You must specify one of either LIN, OCT, or DEC. LIN specifies a linear sweep of the frequency. OCT or DEC specify equally spaced frequency values using a logarithmic scale. <Points> is the number of points for a linear sweep or the number of points per octave (OCT) or decade (DEC) for either logarithmic sweep. Replace <Start_Freq> with the beginning frequency and <End_Freq> with the final frequency value. Both frequency values have to be positive and the final frequency has to exceed the beginning frequency. Notice that the AC command, unlike the DC command, has no LIST form. The <Points> specification occurs before the starting and ending frequency values. For a linear sweep <Points> is the number of points over the frequency range, not the frequency increment. The frequency increment equals the ratio of the difference between the final frequency and the beginning frequency and one less than the number of points. For either the DEC or OCT logarithmic sweep, <Points> sets the number of intervals in the decade or octave. Usually you give <Start_Freq> and <End_Freq> values that fall at the beginning and end of a decade or octave. In this case, the number of frequency values equals one more than the product of the number of decades or octaves and the <Points> value. If you do not use decade or octave starting values, then the sweep starts at <Start_Freq> and each successive frequency value exceeds the previous frequency value by either a factor equal to ten (DEC) or two (OCT) to the power 1/<Points>. SPICE stops the sweep at the last frequency value remaining less than or equal to <End_Freq>. For example, sweeping the frequency from 0.8 to 5.0 kHz with five points per decade and using the AC statement

    .AC DEC 5 0.8k 5.0k

in your netlist, causes SPICE to find the AC circuit solution at frequencies of 0.8, 1.268, 2.010, and 3.185 kHz, because $10^{(1/5)}$ equals 1.58489 and the next value in the sequence 5.050 exceeds 5.0. The AC statement names no source, since all sources having an AC specification operate at the same frequency. Although SPICE allows only one AC statement in a netlist, you also can have a DC statement in the same netlist. The AC simulation uses the DC operating point set by any DC source values and ignores the DC values in the DC statement.

### 2.3.3 The PRINT AC, PLOT AC, and PROBE Statements

The print statement for AC analysis has the same form as for DC analysis, except that AC replaces DC in the statement. The syntax of the AC print statement is

    .PRINT AC <Output_Var_1> ...

Likewise, you can have SPICE plot AC data in the output file using

    .PLOT AC <Output_Var_1> ...

Using the probe statement

    .PROBE <Output_Var_1> ...

| Modifier | Effect |
|----------|--------|
| M | Magnitude (The default) |
| P | Phase |
| R | Real part |
| I | Imaginary part |
| DB | Twenty times log of value |
| G | Time delay (=$-\partial\phi/\partial\omega$) |

Table 2.9 Modifier Letters

makes PSpice produce a DAT file that contains both the AC and any DC data for use by the Probe application. Using either PRINT AC or PLOT AC without an AC statement in the netlist is meaningless, but PSpice does not warn you of the missing analysis statement. With a specific output-variable list in the probe statement, output values available in the Probe application include only those specific values. This limitation controls the size of the DAT file. The DAT file has data for both AC and DC simulations if both a DC and an AC statement appear in the netlist. When you run Probe using the DAT file for the simulation, Probe makes you choose to plot either the DC or the AC simulation data.

The output variables that appear in a PRINT AC or PLOT AC command have modifier suffixes that allow you to specify magnitude, phase, real part, imaginary part, dB value, or group delay for an AC variable. The modifier letters in Table 2.9 behave as suffixes to the usual current or voltage specifications and have the effects shown in the table. For example, to have the AC magnitude and phase of the voltage across resistance $R_3$ appear in an output file, include

.PRINT AC VM(R3) VP(R3)

in your netlist. In this line you can omit the modifier M, because the phasor magnitude is the default effect. As for a PRINT DC statement, a PRINT AC statement produces no output unless there also is an AC statement in the netlist. These AC suffixes are meaningless for variables in a PROBE statement. However, you can use them when telling Probe to plot a variable when you select an AC section. (See also Sec. 2.6.4.)

The restrictions on the number of variables that can appear in PRINT AC and PLOT AC statements are the same as for their DC versions. Do not specify DC and AC in the same PRINT or PLOT statement, but you can have as many different PRINT or PLOT statements as you wish in a netlist.

## Example 2.2 AC Circuit Analysis

Calculate the real and imaginary part of the admittance $Y(j\omega)$ to five significant digits for the circuit in Fig. 2.4 with $\omega$ equal to 1, 5, and 10 rad/s. Check the calculation values using PSpice.

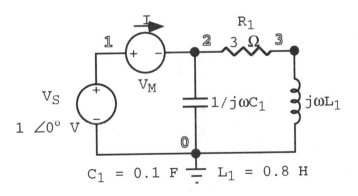

FIGURE 2.4 AC Circuit Analysis Example

## Solution

Using series and parallel rules, the admittance is

$$Y(j\omega) = j\omega C_1 + \frac{1}{3 + j\omega L_1} = j\omega(0.1) + \frac{1}{3 + j\omega(0.8)}$$

The PSpice netlist shown in Fig. 2.5 gives the numerical values of the real and imaginary part of Y(jω) by measuring the current into the admittance using the zero-valued voltmeter VM as an ideal ammeter. The PRINT AC output values have five digits of numerical precision because of the OPTIONS statement. This netlist uses a ten-point linear sweep of the frequency from 1 to 10 Hz, which includes 1, 5, and 10 Hz. The netlist defines parameter "twopi" equal to $2\pi$ numerically by setting twopi = 8*atan(1) in the PARAM statement. However, PSpice does not allow use of parameters in output statements. So this netlist uses a trick to achieve the solution without typing 0.15915 and 1.5915 Hz as the beginning and final frequency values in the AC sweep. The netlist sets the L1 and C1 values smaller than their actual values by factors of $2\pi$, which makes PSpice compute the correct numerical values of $\omega L_1$ and $\omega C_1$. Of course

```
Example 2.2 - AC Circuit Analysis
.OPTIONS NUMDGT=5
.PARAM twopi={8*atan(1)}
Vs 1 0 AC 1
Vm 1 2
R1 2 3 3
L1 3 0 {0.8/twopi}
C1 2 0 {0.1/twopi}
* Playing a game here! With these L1 and C1 values, 1 Hz gives
* the same impedance for L1 and admittance for C1 as 1 rad/s.
* {} notation cannot be used in the .PRINT statement below.
.AC LIN 10 1 10
.PRINT AC IR(Vm) II(Vm)
* Printing the real and imaginary parts of Y.
* With VS a 1-V, 0° source, the current numerically
* equals the admittance.
.END
```

FIGURE 2.5 PSpice Netlist.

| ω rad/s | Re{Y}, S Calc | Im{Y}, S Calc | Re{I}, A PSp | Im{I}, A PSp |
|---------|---------------|---------------|--------------|--------------|
| 1.0000  | 0.31120       | 0.017012      | 0.31120      | 0.017012     |
| 2.0000  |               |               | 0.25952      | 0.061592     |
| 3.0000  |               |               | 0.20325      | 0.13740      |
| 4.0000  |               |               | 0.15593      | 0.23368      |
| 5.0000  | 0.12000       | 0.34000       | 0.12000      | 0.34000      |
| 6.0000  |               |               | 0.093633     | 0.45019      |
| 7.0000  |               |               | 0.074331     | 0.56125      |
| 8.0000  |               |               | 0.060048     | 0.67190      |
| 9.0000  |               |               | 0.049310     | 0.78166      |
| 10.000  | 0.041096      | 0.89041       | 0.041096     | 0.89041      |

Table 2.10 Comparison of PSpice and Computation Y(jω) Values.

PSpice lists the frequency values from the netlist, not the radian frequency values. If you think that this trick is confusing, feel free to edit the netlist, changing the L1 and C1 values to 0.8 and 0.1 and entering the frequency values 0.15915 and 1.5915 Hz instead of 1 and 10 Hz. The tabulation of the calculation values for the real and imaginary parts of $Y(j\omega)$ at 1, 5, and 10 rad/s appears in Table 2.10 for comparison with the PSpice values. You can see that both sets of values are identical.

## 2.4 Transient Simulation

This section describes the transient simulation command TRAN, the transient print and plot commands PRINT TRAN and PLOT TRAN, and the transient independent source specifications. Because transient problems in many cases occur due to switching events, voltage- and current-controlled switch device statements and their model syntax are given here.

### 2.4.1 The TRAN Command

The TRAN command has the syntax

.TRAN [/OP] <t_Print> <t_Final> [<t_No_Print> [<t_Ceil> ]] [UIC]

where the optional control characters have the following effects:

- /OP Causes printing of operating point values.
- UIC Makes PSpice use initial conditions of capacitances and inductances to determine the initial operating point of the circuit.

Table 2.11 lists the time parameters in the TRAN command and describes their meaning, units, and default values. PSpice uses an internal time step that changes as the computation proceeds. For this reason, output variable values generally do not exist for time values differing exactly by <t_Print>, so PSpice computes the values at the print times using a second-order interpolation. You need to use judgment to determine appropriate values for all of these transient time parameters. The value of <t_Print> has to be short enough to resolve variations that occur, but not so short that too much data appear in the output. The value of <t_Final> needs to be long enough to show the complete waveform, without wasting computation time looking at too many steady-state solution values. SPICE always starts transient solutions at time zero and knows nothing about times prior to zero. You may have to time-shift your problem, so that important transient events occur only at or after time zero. If you know that no important transient information exists until a certain time after zero, specify a

| Parameter | Meaning | Units | Default |
|-----------|---------|-------|---------|
| <t_Print> | Time between print values | s | |
| <t_Final> | Final time of simulation. | s | |
| <t_No_Print> | Time before printing | s | 0 |
| <t_Ceil> | Ceiling on numerical time steps | s | <t_Final>/50 |

Table 2.11 Transient Specification Parameters

```
Example 2.3 - Transient source specifications
*
Vdcac        1    0     DC    2      AC   3      180
Vexp         2    0     DC    0.5
+                       EXP(1 5 1 2 4 3)
Vpulse       3    0     DC    1
+                       PULSE(1 2 1 0.05 0.1 0.85 4)
Vsin         4    0     DC    1
+                       SIN(1 2 0.5 1 0.5 45)
Vpwl         5    0     DC    1
+                       PWL[(1,5) (3,-1) (4,1)]
.TRAN        0.2  8
.PRINT TRAN V(1)   V(2)   V(3)   V(4)   V(5)
.PROBE
.END
```

FIGURE 2.6 Netlist to Illustrate Transient Source Specifications

delay before printing time using the <t_No_Print> parameter. If you need to have a ceiling time on the numerical time steps that SPICE uses, specify this value using the <t_Ceil> parameter. Example 2.4 describes a situation where the default value of <t_Ceil> is not appropriate. If you need to specify <t_Ceil>, then you also have to give a value for <t_No_Print>, even if this value is zero.

## Example 2.3 Transient Source Specifications

Write a PSpice netlist to demonstrate the different forms of transient independent voltage source that you can use in PSpice.

### Solution

Figure 2.6 shows a solution for this example. The waveforms at the five nodes demonstrate the DC, EXP SIN, PWL, and PULSE representations of independent voltage sources that you can use in transient simulations. Probe plots of these waveforms appear in Sec. 2.4.2, which describes the syntax of these waveforms and their mathematical forms.

### 2.4.2 Transient Source Specification

The netlist shown in Fig. 2.6 illustrates how to specify independent voltage sources for transient analysis. A complete specification of an independent voltage source has the form

    V<Name> <Node_1> <Node_2> [DC] [<DC_Value>]
    +    [AC <AC_Mag> [<AC_Phase>]]
    +    [<Transient_Specification>]

Current sources have the same form with I replacing V as the first character. The <Transient_Specification> can include one of any of the following

    EXP, SFFM, SIN, PULSE, or PWL

with some or all parameters appropriate to each type.

42

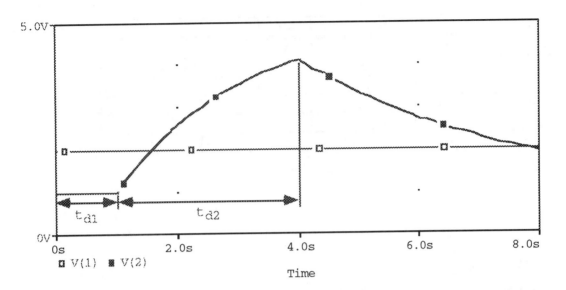

FIGURE 2.7 DC, AC, and Exponential Waveforms

A transient source specification has no meaning for AC analysis. In particular, the SIN transient source does not apply to AC analysis. The DC value of a source having a DC specification and no <Transient_Specification> assumes the DC value during any transient analysis.

### DC and EXP Waveforms

Figure 2.7 shows Probe waveforms of sources Vdcac and Vexp that the netlist in Fig. 2.6 generates. The netlist describes a set of separate circuits, each consisting of a single voltage source that shares only the ground node with the other separate circuits. Source $V_{dcac}$ has no transient specification, so it produces the constant 2-V DC output during the TRAN simulation. The AC specification of the $V_{dcac}$ source has no effect during the transient simulation. SPICE ignores a DC value during a transient simulation for any source having a transient specification.

The form of the EXP source $V_{exp}$ is

$$v_{exp}(t) = \begin{cases} V_1 & t \le t_{d1} \\ V_1 + (V_2 - V_1)\left\{1 - \exp\left[-(t - t_{d1})/\tau_1\right]\right\} & t_{d1} \le t \le t_{d2} \\ V_1 + (V_2 - V_1)\left\{\left(1 - \exp\left[-(t - t_{d1})/\tau_1\right]\right) - \left(1 - \exp\left[-(t - t_{d2})/\tau_2\right]\right)\right\} & t \ge t_{d2} \end{cases}$$

SPICE syntax for the EXP source is

    V<name> <Node_1> <Node_2>
    +      EXP(<V₁> <V₂> <t_d1> <τ₁> <t_d2> <τ₂>)

Table 2.12 gives the EXP source parameters, their meanings, units, and default values. The values of <V₁> and <V₂> can be positive, negative, or zero. Either <V₁> or <V₂> can be the larger value.

| Parameter | Meaning | Units | Default |
|-----------|---------|-------|---------|
| $<V_1>$ | Starting value | V | None |
| $<V_2>$ | Peak value | V | None |
| $<t_{d1}>$ | Rise delay | s | 0 |
| $<\tau_1>$ | Rise time constant | s | $<t\_Step>$ |
| $<t_{d2}>$ | Fall delay | s | $<t_{d1}> + <t\_Step>$ |
| $<\tau_2>$ | Fall time constant | s | $<t\_Step>$ |

Table 2.12 EXP Waveform Parameters

### *Pulse Waveform*

The PULSE source, shown in Fig. 2.8 has the syntax

```
V<name> <Node_1> <Node_2>
+       PULSE(<V1> <V2> <td> <tr> <tf> <tp> <tper>)
```

where Table 2.13 gives the parameters, their meanings, units, and default values. The values of $<V_1>$ and $<V_2>$ can be positive, negative, or zero. Either $<V_1>$ or $<V_2>$ can be the larger value. The delay time $<t_d>$ equals the time before the waveform first changes from $<V_1>$ to $<V_2>$. The rise time $<t_r>$ is the time duration for the waveform to change from $<V_1>$ to $<V_2>$, and the fall time $<t_f>$ is the time duration of the fall from $<V_2>$ to $<V_1>$. The pulse duration $<t_p>$ is the time spent at the $<V_2>$ level for each period. The period $<t_{per}>$ is the time between repetitive points of the waveform.

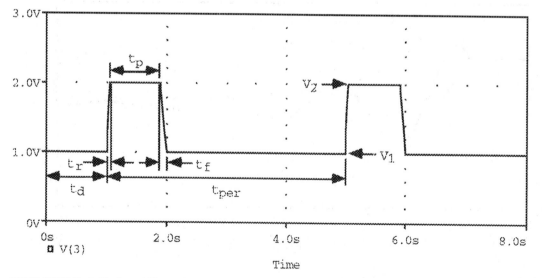

FIGURE 2.8 Pulse Waveform

| Parameter | Meaning | Units | Default |
|-----------|---------|-------|---------|
| <V₁> | Starting value | V | None |
| <V₂> | Pulse value | V | None |
| <t_d> | Time delay | s | 0 |
| <t_r> | Rise time | s | <t_Step> |
| <t_f> | Fall time | s | <t_Step> |
| <t_p> | Pulse duration | s | <t_Final> |
| <t_per> | Period | s | <t_Final> |

Table 2.13 Pulse Waveform Parameters

### Sine-Function Waveform

The SIN waveform, shown in Fig. 2.9, has the equation

$$v_{sin}(t) = \begin{cases} V_1 + V_2 \sin\{2\pi\phi/360\} & t \le t_d \\ V_1 + V_2 \exp[-\alpha(t-t_d)]\sin\{2\pi[f(t-t_d)+\phi/360]\} & t \ge t_d \end{cases}$$

The syntax of the transient sine function is

```
V<name> <Node_1> <Node_2>
+     SIN(<V₁> <V₂> <f> <t_d> <α> <φ>)
```

where Table 2.14 gives the parameters, their meanings, units, and default values. The phase shift $\phi$ is a PSpice feature, not available with SPICE. As for EXP and PULSE waveforms, <V₁> and <V₂> can have any size or polarity.

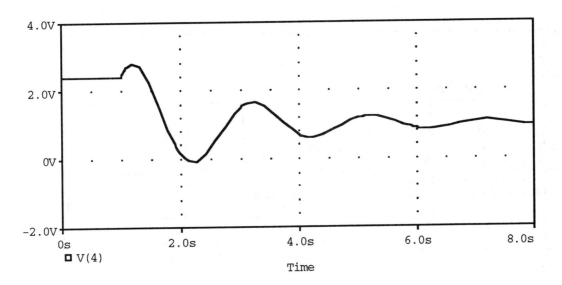

□ V(4)

FIGURE 2.9 Sine-Function Waveform

| Parameter | Meaning | Units | Default |
|-----------|---------|-------|---------|
| $<V_1>$ | Offset value | V | None |
| $<V_2>$ | Amplitude | V | None |
| $<f>$ | Frequency | Hz | 1/<t_Final> |
| $<t_d>$ | Delay time | s | 0 |
| $<\alpha>$ | Damping factor | $s^{-1}$ | 0 |
| $<\phi>$ | Phase | Degrees | 0 |

Table 2.14  Sine-Function Waveform Parameters

### PWL Waveform

The PWL waveform shown in Fig. 2.10 consists of straight-line segments that connect adjacent points. The syntax of this waveform is

V<Name>  <Node_1>  <Node_2> PWL($<t_1>$ $<V_1>$ $<t_2>$ $<V_2>$ ... )

The PWL source has the $<V_1>$ value until time $<t_1`>$ even if $<t_1>$ is nonzero. The last voltage value continues until the end of the simulation, even if the last time is less than the final simulation time. Time values have to increase successively. Using square braces, parentheses, and commas in the netlist helps to delineate successive pairs of time and voltage parameters. For example,

Vpwl 5 0 PWL[(1,5) (3,-1) (4,1)]

is more readable than

Vpwl  5  0  PWL(1 5 3 -1 4 1)

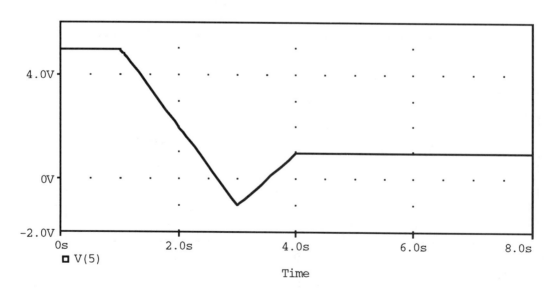

FIGURE 2.10 PWL-Source Waveform

| Parameter | Meaning | Units | Default |
|-----------|---------|-------|---------|
| $<V_1>$ | Offset voltage | V | 0 |
| $<V_2>$ | Voltage amplitude | V | None |
| $<f_c>$ | Carrier frequency | Hz | 1/<t_Final> |
| $<M>$ | Modulation index | | 0 |
| $<f_m>$ | Modulation frequency | Hz | 1/<t_Final> |

Table 2.15  SFFM-Function Waveform Parameters

### SFFM Waveform

A SFFM (Single-Frequency Frequency Modulation) voltage waveform has the syntax

        V<Name> <Node_1> <Node_2>
        +       SFFM(<V_1> <V_2> <f_c> <M> <f_m>)

This syntax describes a time-dependent voltage $v_{SFFM}(t)$, which has the mathematical form

$$v_{SFFM} = V_1 + V_2 \sin\left[2\pi f_c t + M \sin(2\pi f_m t)\right]$$

Table 2.15 lists the parameters for the SFFM waveform, their meanings, units, and default values. Of course, a current source statement begins with I and $<V_1>$ and $<V_2>$ become the offset current $<I_1>$ and current amplitude $<I_2>$, both in amperes.

### 2.4.3 Transient PRINT, PLOT, and PROBE Statements

The transient print statement is similar to the DC print statement, except that TRAN replaces DC. To print transient simulation data in  the output file type

        .PRINT TRAN <Output_1> ...

Select outputs using the same voltage and current variable notation as for DC variable selection. If you wish graphical output  to appear in the output file use

        .PLOT TRAN <Output_1> ...

As for DC and AC simulations, the statement

        .PROBE [<Output_1> ...]

causes PSpice to create an output file having the same name as your netlist, but with a DAT suffix replacing CIR. This file becomes the input file for the Probe application program and lets you create plots of your circuit variables. Omission of <Output_1> ... creates a DAT file that contains all simulation results. Giving specific output variables limits the size of the Probe data file.

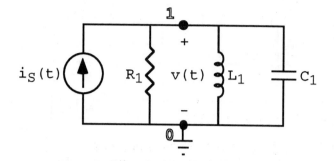

FIGURE 2.11 Parallel RLC Circuit [$i_S(t) = I_{so}u(t)$ with $I_{so}$ = 5 mA, $R_1$ =101/2 kΩ, $L_1$ = 1/(2π) H, $C_1$ = 1/(2π×101) μF]

## Example 2.4 Parallel RLC circuit

For the circuit shown in Fig. 2.11, obtain a plot of the voltage v(t) vs. time using PSpice and Probe. Superimpose the numerical solution on the Probe plot.

### Solution

The PSpice netlist shown in Fig. 2.12 produces the Probe plot that appears in Fig. 2.13. The solution for the circuit in Fig. 2.11 is

$$v_1(t) = \frac{I_{so}}{\omega_n C_1} e^{-\alpha t} \sin(\omega_n t)$$

where

$$\alpha = \frac{1}{2R_1C_1} = \frac{1}{2 \times \frac{101}{2} \times \frac{1}{2\pi \times 101}} = 2\pi \text{ krad/s}$$

$$\omega_n = \sqrt{\frac{1}{L_1C_1} - \alpha^2} = \sqrt{\frac{1}{\frac{1}{2\pi} \times \frac{1}{2\pi \times 101}} - (2\pi)^2} = 20\pi \text{ krad/s}$$

```
Example 2.4 - Parallel RLC circuit
.PARAM twopi={8*atan(1)}
.PARAM Iso=5m Rval={101k/2} Lval={1/twopi} Cval={1u/(twopi*101)}
Is 0 1 {Iso}
R1 1 0 {Rval}
L1 1 0 {Lval}              ;Il1=0
*L1 1 0 {Lval} IC=0        ;An alternate form for preceding line.
C1 1 0 {Cval}             ;Vc1=0
*C1 1 0 {Cval} IC=0        ;An alternate form for preceding line.
.TRAN 10u 1m UIC
*.TRAN 10u 1m 0 5u UIC   ;A change to correct a PSpice problem.
.PRINT TRAN V(1)
.PROBE
.END
```

FIGURE 2.12 PSpice Netlist for Example 2.4.

$$\frac{I_{so}}{\omega_n C_1} = \frac{5}{20\pi \times \dfrac{1}{2\pi \times 101}} = 50.5\,\text{V}$$

Numerically

$$v_1(t) = 50.5\, e^{-6283.2t} \sin(62832t)\,\text{V}$$

This numerical solution appears in Fig. 2.13 with the PSpice solution. Something is wrong! The point of this example is that if you are not careful, PSpice does not obtain a correct solution for a highly oscillatory transient solution. The problem is that the default value of <t_Ceil> is <t_Final>/50. In the netlist for Example 2.4 <t_Ceil> defaults to 20 µs, but the period of the oscillation is $T_n = 2\pi/\omega_n = 100$ µs. This value is too large to give accurate results after many periods. To eliminate this problem, change the TRAN statement as shown by the comment line following the present TRAN statement. The new version makes <t_Ceil> equal to 5 µs, one-twentieth of the oscillation period. With this change PSpice calculates the data that produces the Probe plot shown in Fig. 2.14. The PSpice solution now is consistent with the actual solution. You can run this alternate version of the netlist in Fig. 2.12 by adding an asterisk before the first TRAN statement, making this line into a comment, and removing the asterisk in front of the second TRAN statement. This SPICE problem is unique to transient simulation of a highly underdamped circuit and does not occur for AC simulations of the same circuit. This example shows that <t_Ceil> has to be set to a value at least as small as one-twentieth of the oscillation period. Also, notice that in the example <t_Print> equals one-tenth of the oscillation period. A value at least this small is necessary for the plot to look continuous. Try larger values to see how the solution deteriorates when too few points occur per period.

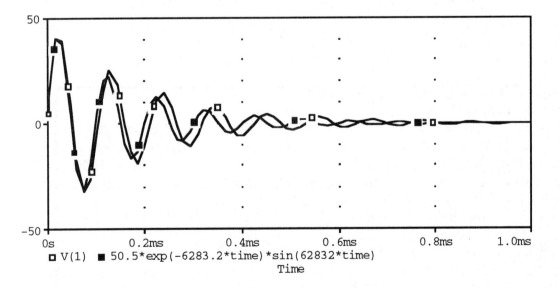

FIGURE 2.13 Probe Plot of v(1) for Example 2.4

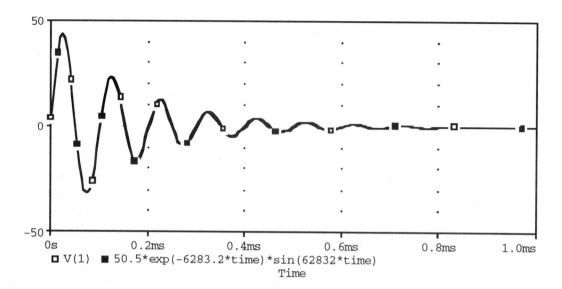
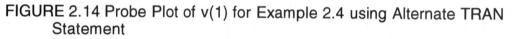

FIGURE 2.14 Probe Plot of v(1) for Example 2.4 using Alternate TRAN Statement

### 2.4.6 Switch   Devices

Transients in circuits often occur because of switching events. In SPICE, switches are voltage- or current-controlled resistances that change exponentially from an off-resistance value to an on-resistance value, as the control variable changes from an off to an on value. The syntax to describe a voltage-controlled switch is

> S<Name> <Plus_Node> <Minus_Node>
> +       <Plus_Control> <Minus_Control> <Model_Name>

<Plus_Node> and <Minus_Node> are the nodes that the switch connects or disconnects. <Plus_Control> and <Minus_Control> are the nodes of the controlling voltage source. <Model_Name> refers to a switch model that you specify with the model statement

> .MODEL <Model_Name> VSWITCH [<Parm_1> ...]

The optional switch parameters [<Parm_1> ...] their meaning, units, and default values appear in Table 2.16. To avoid numerical computation problems, do not change the ratio of ROFF to RON to a value greater than $10^{12}$. Both RON and ROFF should be nonzero but less than 1/GMIN, where the default value of GMIN equals $10^{-12}$ S. GMIN can be set using the OPTIONS command. The value of VON can be less than the value of VOFF if you wish to have the switch on for low control-voltage values. Avoid making the difference between VON and VOFF small. In many situations you use an independent transient voltage source that connects between a node in your circuit, usually the ground node, and an isolated node to control the voltage-controlled switch.

| Parameter | Meaning | Units | Default |
|-----------|---------|-------|---------|
| RON | On resistance | $\Omega$ | 1.0 |
| ROFF | Off resistance | $\Omega$ | $10^6$ |
| VON | On control voltage | V | 1.0 |
| VOFF | Off control voltage | V | 0 |

Table 2.16 VSWITCH Parameters with Default Values

For a current-controlled switch, use

W<Name> <Plus_Node> <Minus_Node> <Control_V_Source>
+       <Model_Name>

where <Model_Name> appears in the model statement

.MODEL <Model_Name> ISWITCH [<Parm_1> ...]

In this model VON and VOFF of Table 2.16 become ION and IOFF, having default values equal to 1 mA and 0 mA, respectively. <Control_V_Source> names the voltage source that measures the controlling current. In many situations you create a separate circuit consisting of an independent transient current source in series with a null-valued voltage source and sharing only a common node with the main circuit to control the current-controlled switch.

## Example  2.5 Switching  Transient

The circuit in Fig. 2.15  illustrates a switching transient. With the circuit in steady-state operation at t = 0 with switch S1 open, the switch closes for 5 s, then opens. Simulate this circuit with PSpice and use Probe to obtain plots of the inductance current $i_L(t)$ and capacitance voltage $v_C(t)$ for $0 \le t \le 10$ s.

### Solution

With switch S1 open and the circuit in steady state at t = 0, the initial inductance current $i_L(0) = 0$ and capacitance voltage $v_C(0) = 5 - 1 = 4$ V. These values are the circuit's initial condition values and they appear in the netlist of Fig. 2.16. When switch S1 closes, the circuit has two parts: an $R_1L$

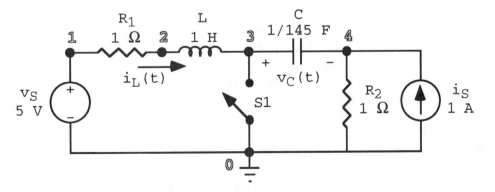

FIGURE 2.15 RLC Switching Circuit Example

```
Example 2.5 - Switching Transient
Vs 1 0 5
* Source Vc changes from 1 V to 0 V in 1 ms at t = 5 s.
Vc 5 0 PWL[(5,1) (5.001,0)]
Is 0 4 1
R1 1 2 1
R2 4 0 1
L  2 3 1              ; Defaults to IC=0
C  3 4 {1/145} IC=4
S1 3 0 5 0 Sxx
.MODEL Sxx VSWITCH RON=1m ROFF=1Meg
.TRAN 25m 10 0 10m UIC
.PRINT TRAN I(L) V(3,4)
.PROBE
.END
```

FIGURE 2.16 Netlist for Example 2.5

circuit having a time constant of 1 s, and an $R_2C$ circuit with a 1/145 s time constant. The switch opens again after 5 s and the circuit becomes a series RLC circuit with the characteristic equation

$$s^2 + \frac{R_1 + R_2}{L}s + \frac{1}{LC} = 0$$

Using the numerical values

$$s^2 + 2s + 145 = 0 \qquad\qquad s = -1 \pm j12 \text{ rad/s}$$

so the circuit solution is oscillatory. The damping factor is 1 Nep/s, so the solution returns to steady-state operation in about 5 s. The natural frequency is 12 rad/s, so the period of oscillation is $\pi/6$ s, which is approximately 500 ms. To display the oscillation with about 20 points per period requires that we set the time step to 25 ms. We set the step-ceiling value to 10 ms to assure that PSpice accurately calculates the waveforms.

For accurate simulation of the circuit, the switch resistance $R_{ON}$ has to be much smaller than either $R_1$ or $R_2$. For this reason, the switch model sets $R_{ON}$ to 1 m$\Omega$. Letting $R_{OFF}$ equal 1 M$\Omega$ approximates an open circuit and makes the dynamic range of the switch resistance $10^9$, which is less than $10^{12}$.

The switch is on for 5 s, then turns off. To have the switch behave in this manner, we let $V_{ON}$ and $V_{OFF}$ have their default values of 1 V and 0 V. Using the PWL function, we let the control voltage $v_C$ be 1 V for the first 5 s, change to 0 V in 1 ms, and stay at that value for the next 5 s. The 1-ms time that we allow for the control-voltage change is nearly instantaneous in comparison to the time that it takes for voltages and currents in the circuit to change significantly, because the damping factor is 1 rad/s with the switch on and the shortest time constant with the switch off is 1/145 s.

The control voltage connects between node 5 (not shown in Fig. 2.15) and node 0. Although control voltage $v_C$ is the only element in the netlist that connects to node 5, PSpice places a resistance value equal to 1/GMIN across the control-voltage terminals, so you do not have to provide a second element connection to node 5.

The netlist includes the UIC option in the TRAN statement so that the initial operating point starts with initial inductance current and capacitance voltage values appropriate to steady-state operation with switch S1 off. If the TRAN statement does not include UIC, then PSpice will calculate the operating point with switch S1 on, since the control voltage $v_C$ is 1 V at t = 0. As an alternate solution to this problem, you can remove the UIC statement and replace the control-source statement with

Vc    5    0    PWL [(0,0) (1m,1) (5,1) (5.001,0)]

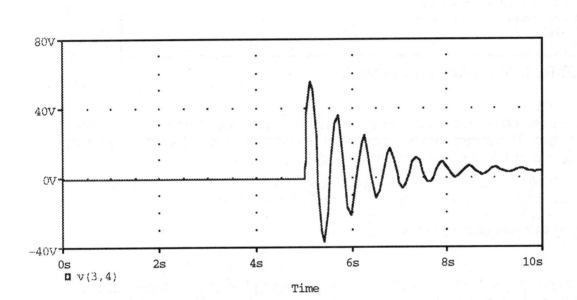

a) Capacitance Voltage Waveform.

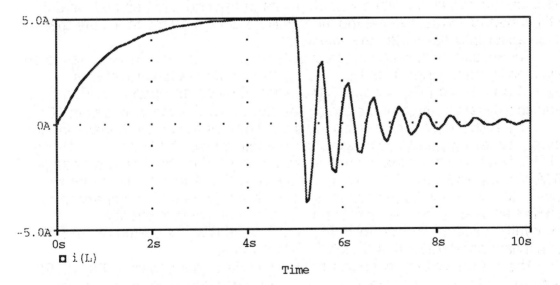

b) Inductance Current Waveform.

FIGURE 2.17 Example 2.5 Waveforms

This version of the control-source statement starts with the switch off, so with no UIC option SPICE finds the operating point with the switch off. Then, within 1 ms, the switch turns on and remains on until 5 s. Finally, during the 1-ms time after 5 s, the switch turns off again, remaining off for the rest of the simulation.

The Probe plots in Fig. 2.17a and b show the transient behavior that we expect. When the switch is on, the capacitance voltage falls from 4 V to -1 V so quickly that this portion of the transient is not observable on the time scale of this plot. You can use **Set_range** on the **X_axis** menu to change the x-axis time scale range to run from 0 to 50 ms to see $v_C(t)$ change exponentially from 4 V to -1 V. The inductance current rises more slowly to the steady-state value of 5 A, since the time constant for this transient is 1 s. After the switch opens at t = 5 s the solution becomes oscillatory. The period of the oscillation is slightly greater than 0.5 s, since two oscillations occur in a time slightly greater than 1 s. The value of the oscillation period we expect to see is $\pi/6 \approx 0.524$ s.

## 2.5 Advanced PSpice Simulation Techniques

The topics in this section describe PSpice commands that significantly enhance circuit analysis capabilities. Some may involve circuit topics that you encounter in second or third courses, or not at all. For this reason, these topics appear here rather than in the earlier sections.

### 2.5.1 Analog Behavioral Modeling

Analog behavioral modeling, a recent innovation in PSpice circuit modeling, is the enhancement of E- and G-type controlled sources to include VALUE, TABLE, FREQ, LAPLACE, and CHEBYSHEV keywords. For completeness, this section also describes the older polynomial form of controlled sources, but the polynomial form is awkward to use. The VALUE form is much easier to use and more general in nature, so is preferable. F and H sources have the older polynomial form but none of the behavioral modeling forms. When using any of the behavioral modeling keywords, remember to type a space after the analog behavioral keyword. Even though VALUE does not require a space, using one helps to prevent forgetting to use a space with the other keywords. In the syntax examples shown below, you do not need to type spaces before and after equality signs, but doing so improves readability of the netlist.

### *POLY Form*

All four types of controlled source (E, F, G, or H) can have a polynomial dependence on their controlling variable. For example, to make an E source have a polynomial dependence, write

```
E<Name> <Node_1> <Node_2> POLY(<Dim>)
+     (<Node_3>, <Node_4>) ... <Coefficients>
```

Choose <Name> to identify the E source with the VCVS that it simulates in the circuit diagram. As usual, <Node_1> and <Node_2> are the nodes that the source connects between. <Node_3> and <Node_4> identify the controlling nodes. The keyword POLY indicates that the dependent source is a polynomial function of the controlling variable(s). <Dim> indicates the number of the controlling variables, telling SPICE how many sets of controlling nodes to expect.  Replace <Coefficients> with the sequence of polynomial coefficients. For example, to create a VCVS source Esample that sets the voltage between node 3 and node 4 to

$$V(3,4) = 2.0 + 3.5 \times V(1,2) + 4.0 \times V(2,0) + 5.0 \times V(1,2) \times V(2,0)$$

type

> Esample 3 4 POLY(2) (1,2) (2,0) 2.0 3.5 4.0 5.0

If any terms are missing from the polynomial, enter a zero coefficient. For example, if the V(2,0) term is missing above, replace 4.0 by 0.0 in the sequence of polynomial values. In the case of F- and H-type sources, the controlling variables have to be voltage sources. These sources cannot be self-referential. For example, do not try to use

> Esample 1 0 POLY(1) (1,0) 1.0 2.0 3.0

You cannot use parameters (see Sec. 2.2.10) to represent coefficient values in this form of dependent source.

### *VALUE Form*

An E-type dependent VCVS source can have  the form

> E<Name> <Node_1> <Node_2> VALUE = {<Expression>}

where <Name>, <Node_1>, and <Node_2> have their usual meanings. A G-type source also has this VALUE form, with the letter G replacing E. The keyword VALUE makes this dependent source assume the value resulting from evaluation of <Expression>, which can contain constants, parameters, voltages using node or node-difference notation, currents in voltage sources, arithmetic operations +, -, *, and /, parentheses to control operator precedence, the variable time (TIME), and any of the functions in Table 2.8. You must type both curly braces {}. Since the E- and G-type controlled sources can depend upon a current, the F- and H-type current-controlled sources are superfluous and do not have the VALUE form. Be careful not to make an E-type source depend on its own node voltage difference.

For DC and bias-point analysis the value of <Expression> applies with TIME equal to zero. During transient analysis the E source assumes successive instantaneous values.

For AC analysis, PSpice uses a linearization of <Expression> at the operating point. To explain the concept of linearization about an operating point, consider  the PSpice statement

> E5 5 0 VALUE = {2*V(1,2)*V(1,2)}

and  assume that V(1,2) equals 5 V at the operating point. The E5 source makes

$$v_5 = 2v_{12}{}^2$$

so the operating-point value of $v_5$ is

$$v_5\big|_Q = V_5 = 2v_{12}{}^2\big|_Q = 2(5)^2 = 50\,\text{V}$$

where Q denotes operating point. Writing $v_5$ using a Taylor series expansion

$$v_5 = v_5\big|_Q + \frac{\partial v_5}{\partial v_{12}}\bigg|_Q (v_{12} - V_{12}) + \frac{1}{2!}\frac{\partial^2 v_5}{\partial v_{12}{}^2}\bigg|_Q (v_{12} - V_{12})^2 + \dots$$

Neglecting the second and higher-order terms

$$v_5 - V_5 \approx \left\{ 4v_{12}\big|_Q \right\}(v_{12} - V_{12}) = 20(v_{12} - V_{12})$$

For this example, in AC analysis PSpice makes node voltage 5 depend upon the node-voltage difference between node 1 and node 2 with a proportional constant of 20 V/V.

You can easily have AC source values that cause the change of circuit variables to exceed the conditions for accurate characterization of the circuit by linearization about an operating point. PSpice does not tell you when linearization is not valid. However, the motivation for using AC analysis with nonlinear circuits is to determine transfer functions, which are ratios of response variables to excitations. The transfer function found using small-signal analysis is valid even when the size of the signals cause the conditions of small-signal analysis not to be valid. Using the transfer function under these conditions causes errors, but that is your problem, not PSpice's!

## TABLE Form

The TABLE form of E-type dependent source has the syntax

```
E<Name> <Node_1> <Node_2> TABLE  {<Expression>}
+      = (<Input_1>, <Output_1>), ... , (<Input_n>, <Output_n>)
```

Here the keyword TABLE indicates that the source assumes the value set by interpolation of the table of pairs of values using the result of <Expression> as the input value. When the value of  <Expression> is less than <Input_1> the value of the dependent source equals <Output_1>, and if <Expression> is greater than <Input_n> the value of the dependent source equals <Output_n>. Successive input values in the table have to increase monotonically. You do not have to type the parentheses and commas, but doing so helps to delineate the pairs of values. G sources also can use the TABLE keyword.

## FREQ Form

Both E and G sources can use the FREQ keyword. For an E source the general form with FREQ is

```
E<Name> <Node_1> <Node_2> FREQ {<Expression>}
+      = (<Freq_1>, <Mag_1>, <Phase_1>), ... ,
         (<Freq_n>, <Mag_n>, <Phase_n>)
```

The data describes the frequency response of a transfer function. Each set of data following  {<Expression>} has a frequency in Hz, a magnitude in dB, and a phase in degrees. For frequency values between <Freq_1> and <Freq_n> PSpice interpolates phase linearly and magnitude logarithmically. With frequency values less than <Freq_1>, <Mag_1> and <Phase_1> are the values of the transfer function. With frequency values greater than <Freq_n>, <Mag_n> and <Phase_n> are the values of the transfer function.

For DC and bias-point analysis the value of the E source is the value of <Expression> times the table value. For AC analysis the magnitude is the product of the magnitude of <Expression> times the magnitude of the dB table value, and the phase is the sum of the phase of <Expression> and the phase table value. For transient analysis PSpice interprets the frequency response as a Fourier transform.

## LAPLACE Form

The Laplace form of an E or G source expresses the source response in terms of a Laplace transform, using the Laplace variable s and an expression. For an E source, the Laplace statement has the form

```
E<Name> <Node_1> <Node_2> LAPLACE {<Expression>}
+       = {<Laplace_Transform>}
```

<Name>, <Node_1>, <Node_2>, and {<Expression>} have the same significance as for the syntax of the other behavioral modeling keywords. Replace <Laplace_Transform> with an expression involving the complex frequency s. You can use arithmetic operations +, -, *, and /, parentheses, and any of the functions in Table 2.8.

For DC and operating point analysis the E device has a value that equals the value of <Expression> times the value of <Laplace_Transform> with s equal to 0. AC analysis linearizes <Expression> about the operating point, then determines the magnitude and phase of the E device by taking the complex number product of <Expression> and <Laplace_Transform>, with s = jω in the <Laplace_Transform> expression. Transient analysis makes the E source value at any time equal to the convolution of the instantaneous values of <Expression> with the inverse transform of <Laplace_Transform>.

## 2.5.2 Subcircuit Statement

When a circuit repeats instances of the same circuit structure, you can represent this subcircuit with a single set of PSpice statements. PSpice provides this capability with the SUBCKT and ENDS statements. When a circuit does not have multiple instances of a subcircuit, encapsulation of a subcircuit using these statements may be conceptually useful. Even when the subcircuit contains only a single element, using a subcircuit may be useful. When you have a subcircuit description in a netlist, PSpice expands the circuit description before doing the various analysis processes. In effect, PSpice works with an effective netlist that repeats the subcircuit description everywhere your original netlist makes a subcircuit call.

To install a subcircuit in your netlist, type

```
X<Name> <Node_1> <Node_2> ... <Node_n> <Subckt_Name>
+       [PARAMS: <Par_Name_1 > = <Call_Value_1> ...]
```

Choose <Name> to relate a specific instance of a subcircuit to the appearance of this subcircuit in your circuit. For example, if you have an op amp OA1 in your circuit, you might use OA1 to name the subcircuit in your netlist. Replace

<Node_1>, <Node_2>, ... <Node_n> with the node numbers of the subcircuit in your circuit. The number of these nodes has to agree with the number of nodes in the subcircuit description. <Subckt_Name> relates the subcircuit statement to the subcircuit model. Using the PARAMS: keyword lets you pass parameters or values to the subcircuit. This ability means that the subcircuit does not have to have the same parameter values each time that you use it.

A subcircuit definition has the form

```
.SUBCKT <Subckt_Name> <Node_1> <Node_2> ... <Node_n>
+      [OPTIONAL: <Interface_Node_1> <Default_Value_1> ...]
+      [PARAMS: <Parameter_Name_1> = <Value_1> ...]
...

...
.ENDS
```

Between the SUBCKT line and the ENDS line, any circuit description line can appear, e.g., R, L, C, V, I, or E. No command lines (lines beginning with a period) can appear within the subcircuit, except for a MODEL statement. A MODEL statement within a subcircuit exists locally, and lines outside of the subcircuit cannot refer to a model having a definition within a subcircuit. However, a subcircuit device can use a model in the main netlist.

Comments or comment lines within a subcircuit are very useful. Usually a comment or set of comments precede a subcircuit description to describe the function of the circuit, and these help to identify the significance of each node in the SUBCKT statement.

Node names (except for global node names), device names, and model names in device statements within the subcircuit definition are local to the subcircuit. For example, node 5 within the subcircuit is unknown outside of the subcircuit and does not conflict with node 5 in the main part of the netlist. The main circuit shares ground node 0 with the subcircuit. Global nodes, using the notation $G_<Node> are available within subcircuits. Using OPTIONAL: lets you identify optional subcircuit nodes in the X statement. When the X statement does not use an optional node, then the default node applies. When an X statement refers to an optional node, all preceding optional nodes have to be given.

You can call a subcircuit within a subcircuit, but you may not define a subcircuit within a subcircuit. That is, you may include an X-device statement within a subcircuit but not another SUBCKT statement.

If PARAMS: appears in the SUBCKT line then parameter <Par_Name_1 >, having the default value <Value_1> becomes available within the subcircuit to define device values. If the X-statement line includes

PARAMS:  <Par_Name_1 > = <Call_Value_1>

then <Call_Value_1> overrides <Value_1> for this instance of the subcircuit.

## Example  2.6 SUBCKT and the VALUE Form of E Source

Use a subcircuit description to define an op amp having a default voltage gain of $10^5$ V/V with an output voltage limit within 1 V of the positive and negative supply voltages and a differential input resistance with a default value of 2 M$\Omega$.

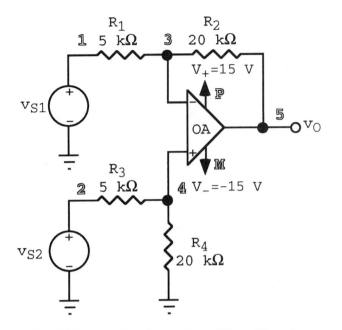

FIGURE 2.18 Op-Amp Amplifier Circuit

```
Example 2.6 - SUBCKT and the VALUE form of E Source
**********    An op-amp model named "Op_Amp" *********
* Default parameters:
*     Ag=10^5, Rin=2MEG
*
* Nodes:          +    -   Output  V+ V-
*                 |    |    |      |  |
.SUBCKT Op_Amp Non Inv  Out    PL ML
+      PARAMS: Ag = 100k, Rin = 2MEG
Rin  Non Inv {Rin}
Eoa  Out 0    VALUE={LIMIT(Ag*V(Non,Inv),V(ML)+1,V(PL)-1)}
*Eoa  Out 0    TABLE {Ag*V(Non,Inv)} = (-14, -14) ( 14,  14)
.ENDS
**********   End of Op-Amp model         *********
* Start of Circuit Description
V+   P   0    15
V-   M   0    -15
*
Vs1  1   0    0
Vs2  2   0    -4
R1   1   3    5k
R2   3   5    20k
R3   2   4    5k
R4   4   0    20k
Xoa  4 3 5 P M  Op_Amp PARAMS: Ag = 10k
.DC   LIN Vs1 -8 8 0.1
.STEP LIN Vs2 -4 4 4
.PRINT DC V(5)
.PROBE
.END
```

FIGURE 2.19 PSpice Netlist

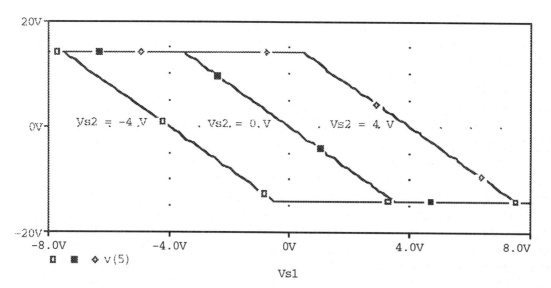

FIGURE 2.20 Probe Plot of $v_O$ vs. $v_{S1}$ with $v_{S2}$ Equal to -4, 0, and 4 V

Use this subcircuit in the simulation of the circuit shown in Fig. 2.18 to display the output voltage $v_O$ vs. $v_{S1}$ with $v_{S1}$ varying from -8 V to 8 V and $v_{S2}$ equal to -4, 0, and 4 V using Probe. In the simulation set the op-amp gain to $10^4$ V/V, leaving the input and output resistances at their default values.

### Solution

The netlist in Fig. 2.19 satisfies the specifications. Alphanumeric node labels in the SUBCKT description help to document the node definitions. The comment line preceding the SUBCKT statement reinforces this documentation. The limit

function makes the E source have a voltage gain of $A_g$ for values between $V_- + 1$ and $V_+ - 1$. If you replace the nodes PL and ML in the subcircuit with P and M, the simulation has the same result because nodes within subcircuits are local to the subcircuit. However, this netlist is more readable with different names for the local power supply nodes and the main circuit's supply nodes. The Eoa source that appears as a comment is an alternate TABLE form. With this statement replacing the preceding statement, you can eliminate reference to the positive and negative supply voltage terminals PL and ML. However, this alternate solution does not allow for the effect that variation of the op-amp supply voltage has on the op-amp limit voltage. The Xoa subcircuit statement shows that the subcircuit call can override a subcircuit default parameter, in this case $A_g$. The Probe plot of $v_O$ vs. $v_{S1}$ with $v_{S2}$ as a parameter appears in Fig. 2.20. This plot shows an output voltage limiting effect typical of op-amp circuits.

### 2.5.3 Library Statement

As you develop a set of subcircuits, device models, or useful parameters that you use frequently, you can store these for convenience in a separate file. The evaluation version of PSpice has a file "EVAL.LIB" that contains a few op amps, comparators, diodes, bipolar transistors, and MOSFET models. To access these subcircuits and models from your netlist, tell PSpice where to find them using the LIB statement

.LIB [<File_Name>]

where <File_Name> names a file that PSpice is to search. If you omit <File_Name> then PSpice assumes the appropriate library is NOM.LIB. The NOM.LIB library does not exist for the evaluation version. The evaluation version library is EVAL.LIB, so you have to provide this name in the LIB statement. If your file name has a suffix, such as LIB, type the suffix since PSpice does not add LIB by default. PSpice creates and associates an index file with each library file. For example, once you access the EVAL.LIB file your PSpice folder will contain a file with the name EVAL.IND. The index file helps PSpice locate models and subcircuits quickly. When your netlist refers to a model, subcircuit, or parameter not in the netlist, PSpice searches the current working directory or folder for <File_Name> to find the model, subcircuit, or parameter. PSpice creates a working netlist by embedding the reference model, subcircuit, or parameter, using only those references that it needs. You can use more than one LIB statement in a netlist, so you can organize your files any way that you like. Library files can contain comment statements, MODEL statements, PARAM statements, and subcircuit definitions that begin with a SUBCKT statement, and end with an ENDS statement.

### 2.5.4 Function and Include Statements

The FUNC statement allows you to define functions using any of the arithmetic operators +, -, *, and /, parentheses, or standard functions from Table 2.8. The statement defining a function has to appear before any reference that uses the function. The function definition statement has the form

.FUNC <Fct_Name>([<Arg_1>, …]) <Fct_Def>

The name of the function <Fct_Name> may not use any of the function names in Table 2.8. A function can have no arguments or as many as ten arguments. In any case, the parentheses have to appear. A reference to a function has to contain as many arguments as appear in the definition. The definition of a function <Fct_Def> has to fit on one line, but it may refer to previous function definitions. The Laplace variable s and parameters can appear in the function definition. Do not use in-line comments after a function definition. PSpice expands the function, replacing arguments by their actual values and surrounding the function definition with parentheses.

PSpice inserts the contents of another file into your file if you use the INC statement. This statement, having the form

.INC "<File_Name>"

physically inserts the contents of <File_Name> into your netlist. The include file cannot have a title line. An END statement appearing in the include file only marks the end of the include file. You can nest include statements to a depth of four levels.

### 2.5.5 Monte Carlo, Sensitivity, and Worst-Case Analysis

The Monte Carlo (MC), sensitivity (SENS), and worst-case (WCASE) commands give statistical and variational information about a circuit's performance.

### *Monte Carlo Simulation*

To do a Monte Carlo simulation, type

        .MC <No_Runs> <Analysis> <Output_1> ...
        +       <Function> [<Option> ...] [SEED = <Value>]

Specify the number of runs with <No_Runs>, which should not exceed 2000 for printed results or 100 for Probe simulation. Replace <Analysis> with one of either DC, AC, or TRAN. The output list <Output_1> ... tells SPICE which output variables to record. Use the same format here as for the corresponding PRINT statement. The <Function> argument tells PSpice how to reduce the variations to a single value. Replace <Function> with one of the following:

> YMAX – Finds the greatest difference for each run from the nominal simulation

> MAX – Finds the maximum of each simulation

> MIN – Finds the minimum of each simulation

> RISE_EDGE(<Value>) – Finds the first occurrence of the simulation above <Value>

> FALL_EDGE(<Value>) – Finds the first occurrence of the simulation below <Value>

If you specify <Option> it can be any or none of the following:

> LIST – Prints parameter values of components for each run

> OUTPUT <Type> – Asks for output after the nominal run

> ALL – Output for all runs

> FIRST <Value> – Output for first <Value> run

> EVERY <Value> – Output for every <Value> run

> RUNS <Value> ... – Output for runs in the list <Value> ...

> RANGE (<Low>, <High>) – Restricts the range for <Function> evaluation. Use * in place of <Low> or <High> to indicate "for all values."

With the optional argument SEED=<Value> you can set the seed number for the random number generator to any odd integer value between 1 and 32,767. The default for the seed value is 17,533.

During a Monte Carlo simulation, the parameters that vary statistically have a uniform distribution by default. Using the OPTIONS command you can change the distribution to Gaussian using

        .OPTIONS DISTRIBUTION GAUSS

To create your own distribution, use the DISTRIBUTION command, which has the syntax

.DISTRIBUTION    <Name>    (<Dev_1>, <Prob_1>, ...)

<Name> is the name that you give to this distribution, and <Dev_1> and <Prob_1> are deviation-probability pairs. These pairs are corner points in a piecewise-linear distribution. Each deviation has to be in the range from -1 to 1 and must equal or exceed the previous value. Each probability has to be greater than or equal to zero. Because PSpice normalizes the probability distribution, the values can exceed one and are relative to each other in size. The maximum number of deviation-probability pairs is 100. PSpice uses this distribution in the following four-step procedure:

- Generate a random number $n_R$ between 0 and 1
- Normalize the area of the probability distribution to 1
- Set the random deviation value to the deviation $x_D$ that makes the area between -1 and $x_D$ equal to the random number
- Apply this deviation to compute a model parameter

For example, consider the bimodal probability distribution

.DISTRIBUTION bimodal  (-1,1)  (0,0)   (1,1)

The area of this distribution equals 1, so already is normal. The value of the deviation $x_D$ in this case is

$$n_D = \begin{cases} -\sqrt{1 -- 2n_R} & n_R \le 0.5 \\ \sqrt{2n_R - 1} & n_R > 0.5 \end{cases}$$

For the value $n_R$ that the random number generator gives, the model parameter $x_M$ that PSpice uses for the nominal model parameter $x_N$ with tolerance $x_T$ is

$$x_M = x_N + n_D x_T$$

If the tolerance is a percent, replace $x_T$ with the percent value times the nominal value and divide by 100.

### *Sensitivity   Analysis*

Sensitivity analysis measures the change of a specific output with respect to various input values or circuit parameters. The completely normalized sensitivity of an output y to a parameter x is

$$S_x^y = \frac{\partial y / y}{\partial x / x} = \frac{x}{y} \times \frac{\partial y}{\partial x} = \frac{\partial \ln(y)}{\partial \ln(x)}$$

This completely normalized sensitivity measures the percentage change of the response to the percentage change of the parameter x. PSpice determines the sensitivity when you place the statement

.SENS <Output_Var1> ...

in the netlist. This statement causes PSpice to linearize the circuit about the operating point and find the change of the output to a small change of the input. PSpice then lists the unnormalized sensitivity

$$S_x^y = \frac{\partial y}{\partial x}$$

and the sensitivity partially normalized with respect to the x parameter

$$S_x'^y = \frac{\partial y}{\partial x / x}$$

Then

$$S_x'^y = x S_x^y$$

## Example 2.7 Sensitivity Analysis

Determine the sensitivities of the output voltage to all parameters and the four voltage sources for the R-2R ladder circuit shown in Fig. 2.21.

### Solution

The netlist in Fig. 2.22 gives a PSpice solution for this example. The nominal output voltage is

$$V_{Out} = \frac{1}{8} V_{S0} + \frac{1}{4} V_{S1} + \frac{1}{2} V_{S2} + V_{S3}$$

You can derive this result by successively collapsing sources $v_{S0}$, $v_{S1}$, $v_{S2}$, and $v_{S3}$ into Thévenin equivalent circuits. From this nominal solution the completely unnormalized sensitivities of the output voltage to the four sources equal 0.125, 0.25, 0.5, and 1 respectively. These values agree with the unnormalized sensitivities to the source voltages in Fig. 2.23. The partial normalizations also agree with these values (multiply each unnormalized sensitivity by the value of

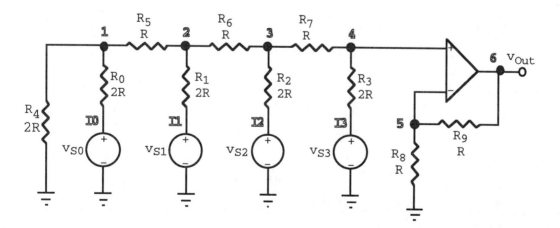

FIGURE 2.21 R-2R Ladder Circuit ($v_{S0}$ = $v_{S1}$ = $v_{S2}$ = $v_{S3}$ = 2 V, R = 5 kΩ)

```
Example 2.7 - Sensitivity Analysis
*
.PARAM Rval=5k
Vs0 I0 0 2
Vs1 I1 0 2
Vs2 I2 0 2
Vs3 I3 0 2
Eoa Out 0 4 5 100k
R0   I0 1 {2*Rval}
R1   I1 2 {2*Rval}
R2   I2 3 {2*Rval}
R3   I3 4 {2*Rval}
R4    1 0 {2*Rval}
R5    1 2 {Rval}
R6    2 3 {Rval}
R7    3 4 {Rval}
R8    5 0 {Rval}
R9    5 Out {Rval}
.SENS V([Out])
.END
```

FIGURE 2.22 Netlist for Sensitivity Analysis

the source and divide by 100 to convert to V/%). A derivation of any of the other sensitivities is quite difficult using normal circuit analysis.

DC SENSITIVITIES OF OUTPUT V(Out)

| ELEMENT NAME | ELEMENT VALUE | ELEMENT SENSITIVITY (VOLTS/UNIT) | NORMALIZED SENSITIVITY (VOLTS/PERCENT) |
|---|---|---|---|
| R0 | 1.000E+04 | -8.398E-06 | -8.398E-04 |
| R1 | 1.000E+04 | -8.594E-06 | -8.594E-04 |
| R2 | 1.000E+04 | -9.375E-06 | -9.375E-04 |
| R3 | 1.000E+04 | -1.250E-05 | -1.250E-03 |
| R4 | 1.000E+04 | 1.660E-05 | 1.660E-03 |
| R5 | 5.000E+03 | 1.641E-05 | 8.203E-04 |
| R6 | 5.000E+03 | 1.562E-05 | 7.812E-04 |
| R7 | 5.000E+03 | 1.250E-05 | 6.250E-04 |
| R8 | 5.000E+03 | -3.750E-04 | -1.875E-02 |
| R9 | 5.000E+03 | 3.750E-04 | 1.875E-02 |
| VS0 | 2.000E+00 | 1.250E-01 | 2.500E-03 |
| VS1 | 2.000E+00 | 2.500E-01 | 5.000E-03 |
| VS2 | 2.000E+00 | 5.000E-01 | 1.000E-02 |
| VS3 | 2.000E+00 | 1.000E+00 | 2.000E-02 |

FIGURE 2.23 Sensitivity Values

### *Worst-Case Analysis*

The worst-case statement

        .WCASE <Analysis> <Output_Var> <Function>
        +       [<Option_1> ...]

varies only one parameter per run and evaluates <Function> for <Output_Var>. <Analysis> can be one of either DC, AC, or TRAN. <Function>, as for MC, can be either YMAX, MAX, MIN, RISE_EDGE, or FALL_EDGE. The possible options that you can use include LIST, OUTPUT ALL, RANGE as for the MC command. In addition, you can specify any of the following:

> HI or LOW – These tell PSpice whether to go high or low relative to the nominal value.

> VARY DEV/LOT/BOTH – Choosing one of DEV, LOT, or BOTH, this option eliminates variation for devices with models not having the selection.

> DEVICES – This option allows you to select the devices to include in the variation. Specify which devices to include by typing a contiguous string using the character for each device – R for resistance, C for capacitance, etc. For example, type "DEVICES RL" to have variation only for resistances and inductances.

You cannot run both MC and WCASE in the same simulation.

### 2.5.6 Fourier Analysis

The FOUR command does a Fourier analysis of one or more transient waveforms for a period T equal to the reciprocal of a given frequency value and commencing at one period prior to the end of the transient simulation. The syntax of the FOUR command is

        .FOUR <Frequency> [<No_Harmonics>] <Output_Var1> ...

Specify the frequency value <Frequency> in Hz. If your netlist includes a FOUR statement it must also have a TRAN statement. The final time <t_Final> in the TRAN statement has to equal or exceed the reciprocal of the frequency value. Specify the number of harmonics that you wish, if different from nine. Indicate the output variables <Output_Var1> ... for which you wish to obtain the Fourier components. PSpice determines the Fourier series in amplitude-phase form using a sine-function rather than a cosine-function form. To compare the phase values that PSpice obtains, remember that

$$\cos(\omega t + \phi) = \sin(\omega t + \phi + 90°)$$

so add 90° to the cosine-function phase value to obtain the sine-function phase value. Be sure that your waveforms are in periodic steady-state operation for the last period of the simulation so that you obtain correct values. PSpice assumes that the waveform existing for that list time period continues periodically. The output that the FOUR command creates includes the *total harmonic distortion* (THD), which is

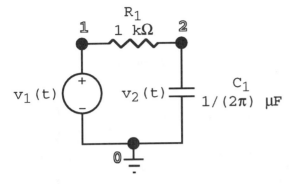

FIGURE 2.24 Circuit Example of Fourier Analysis

$$\text{THD} = \frac{\sqrt{\sum_{n=2}^{n=9} F_n^2}}{F_1} \times 100$$

where $F_n$ is the nth Fourier amplitude of the periodic waveform.

## Example  2.8 Fourier  Analysis

Find the DC value and the first nine Fourier coefficients for the source voltage $v_1(t)$ and the output voltage $v_2(t)$ for the circuit shown in Fig. 2.24, if the source voltage waveform $v_1(t)$ is a 14-V, 1-kHz cosine function that limits at ±14 V. This waveform is a limited-amplitude cosine function, which can occur at the output of an op amp.

### Solution

We use the netlist in Fig. 2.25 to solve this example. The transient simulation runs for 5 ms, which equals five periods of the cosine function. Since the time constant of the circuit equals $1/(2\pi)$ ms, steady-state operation occurs within about 1 ms, and the last 1 ms of the simulation represents periodic steady-state operation for $v_2(t)$. Of course $v_1(t)$ is in steady-state operation immediately. Figure 2.26 shows the Fourier coefficients that PSpice gives for $v_1(t)$ and $v_2(t)$. The normalized component values are equal to the individual frequency component divided by the fundamental component value. The normalized

```
Example 2.8 - Fourier Analysis
*
.PARAM twopi={8*atan(1)}
E1 1 0 VALUE = {limit(18*cos(twopi*1k*time), -14, 14)}
R1 1 2 1k
C1 2 0 {1u/twopi}
.TRAN 10u 6m 5m 2u
.PRINT TRAN V(1) V(2)
.FOUR 1k V(1) V(2)
.PROBE
.END
```

FIGURE 2.25 Netlist for Fourier Example 2.8

```
FOURIER COMPONENTS OF TRANSIENT RESPONSE V(1)

DC COMPONENT =  -4.851417E-07

HARMONIC   FREQUENCY    FOURIER     NORMALIZED    PHASE      NORMALIZED
   NO        (HZ)      COMPONENT    COMPONENT     (DEG)      PHASE (DEG)

    1      1.000E+03   1.581E+01    1.000E+00   9.000E+01    0.000E+00
    2      2.000E+03   1.400E-06    8.850E-08   1.356E+02    4.563E+01
    3      3.000E+03   1.476E+00    9.332E-02  -9.000E+01   -1.800E+02
    4      4.000E+03   1.120E-06    7.084E-08  -4.646E+01   -1.365E+02
    5      5.000E+03   5.433E-01    3.435E-02  -9.000E+01   -1.800E+02
    6      6.000E+03   4.507E-07    2.850E-08   8.903E+01   -9.697E-01
    7      7.000E+03   6.696E-02    4.235E-03   9.000E+01    9.510E-04
    8      8.000E+03   3.695E-07    2.337E-08  -1.424E+02   -2.324E+02
    9      9.000E+03   1.823E-01    1.153E-02   9.000E+01   -1.753E-04

    TOTAL HARMONIC DISTORTION =    1.001936E+01 PERCENT
```

a) Fourier Components of $v_1(t)$

```
FOURIER COMPONENTS OF TRANSIENT RESPONSE V(2)

DC COMPONENT =  -2.984295E-06

HARMONIC   FREQUENCY    FOURIER     NORMALIZED    PHASE      NORMALIZED
   NO        (HZ)      COMPONENT    COMPONENT     (DEG)      PHASE (DEG)

    1      1.000E+03   1.118E+01    1.000E+00   4.500E+01    0.000E+00
    2      2.000E+03   4.851E-06    4.338E-07   1.593E+02    1.143E+02
    3      3.000E+03   4.664E-01    4.171E-02  -1.616E+02   -2.066E+02
    4      4.000E+03   3.029E-06    2.709E-07   1.433E+02    9.833E+01
    5      5.000E+03   1.064E-01    9.512E-03  -1.687E+02   -2.137E+02
    6      6.000E+03   3.393E-06    3.035E-07   1.196E+02    7.459E+01
    7      7.000E+03   9.384E-03    8.393E-04   8.159E+00   -3.684E+01
    8      8.000E+03   3.084E-06    2.758E-07   1.183E+02    7.334E+01
    9      9.000E+03   1.999E-02    1.788E-03   6.342E+00   -3.866E+01

    TOTAL HARMONIC DISTORTION =    4.282917E+00 PERCENT
```

b) Fourier Components of $v_2(t)$.

FIGURE 2.26 Results of Fourier Analysis

phase values are the individual phase values minus the fundamental phase value. The normalized component values let you see at a glance how rapidly the higher harmonics decrease. You can compare normalized phase values directly with values that you compute for a cosine-function form of amplitude-phase series, without changing the phase by 90°.

### 2.5.7 Mutual Inductance

Most circuit theory books describe mutual coupling of two or more inductances in later chapters. For this reason mutual coupling appears here rather than in the sections describing AC or transient analysis. To show that two or more inductances have mutual coupling, include

K<Name> <L_Name1> <L_Name2> ... <Value>

in your netlist. Although SPICE allows coupling of only two inductances, you can couple three or more in PSpice if they have the same coupling coefficient and share a common dot convention. PSpice assumes that the first node of each inductance is the node having the dot, indicating the polarity of the mutual coupling. To use the mutual coupling statement with three or more coils, all coils have to share the same set of dots. In the new notation you can write

```
L1    1    2    5m
L2    3    4    10m
L3    5    6    20m
K123  L1   L2   L3   1
```

to have $L_1$, $L_2$, and $L_3$ have mutual coupling with k = 1. In the old notation, the same effect results using

```
L1    1    2    5m
L2    3    4    10m
L3    5    6    20m
K12   L1   L2   1
K13   L1   L3 1 1
K23   L2   L3 1 1
```

Since Version 5.3 of PSpice, the coefficient of mutual coupling k that <Value> denotes can be between -1 and 1, including 0. Older versions of PSpice require that k have a value greater than 0 and less than 1.

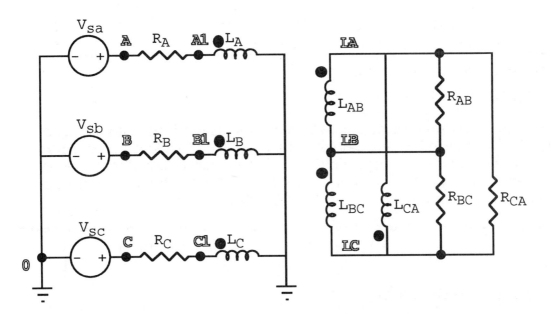

FIGURE 2.27 Three-Phase Power Circuit ($R_A$ = $R_B$ = $R_C$ = 1 Ω, $R_{AB}$ = $R_{BC}$ = $R_{CA}$ = 10 Ω, $L_A$ = $L_B$ = $L_C$ = 10 H, $L_{AB}$ = $L_{BC}$ = $L_{CA}$ = 2.5 H, $K_{AAB}$ = $K_{BBC}$ = $K_{CCA}$ = 1, $V_{sa}$ = 240∠0° V, $V_{sb}$ = 240∠–120° V, $V_{sc}$ = 240 ∠120° V)

## Example 2.9 Mutual Coupling in Three-Phase Power Circuit

Determine the line voltages, phase voltages, line currents, and phase currents on the primary (source) and secondary (load) side of the 60-Hz, power

```
Example 2.9 - Mutual-Coupling in Three-Phase Power Circuit
*
.OPT  NOBIAS
* Sources
Vsa A 0 AC 240      0
Vsb B 0 AC 240  -120
Vsc C 0 AC 240   120
* Primary Circuit
RA  A   A1   1
RB  B   B1   1
RC  C   C1   1
LA  A1  0   10
LB  B1  0   10
LC  C1  0   10
* Secondary Circuit
RAB   LA   LB   10
RBC   LB   LC   10
RCA   LC   LA   10
LAB   LA   LA1  2.5
LBC   LB   LB1  2.5
LCA   LC   LC1  2.5
RABX  LA1  LB   10m
RBCX  LB1  LC   10m
RCAX  LC1  LA   10m
RAN   LA   0    10k
RBN   LB   0    10k
RCN   LC   0    10k
* Primary-to-Secondary Coupling
KAAB LA LAB 1
KBBC LB LBC 1
KCCA LC LCA 1
*
.AC LIN 1 60 60
* Primary line-to-neutral voltages
.PRINT AC V([A1]) VP([A1]) V([B1]) VP([B1]) V([C1]) VP([C1])
* Primary line voltages
.PRINT AC V([A1],[B1]) VP([A1],[B1])
+        V([B1],[C1]) VP([B1],[C1])
+        V([C1],[A1]) VP([C1],[A1])
* Primary line currents
.PRINT AC I(RA) IP(RA) I(RB) IP(RB) I(RC) IP(RC)
* Secondary line-to-neutral voltages
.PRINT AC V([LA]) VP([LA]) V([LB]) VP([LB]) V([LC]) VP([LC])
* Secondary line voltages
.PRINT AC V([LA],[LB]) VP([LA],[LB])
+        V([LB],[LC]) VP([LA],[LC])
+        V([LC],[LA]) VP([LC],[LA])
* Secondary line currents
.PRINT AC I(RAB) IP(RAB) I(RBC) IP(RBC) I(RCA) IP(RCA)
.END
```

FIGURE 2.28 Netlist for Three-Phase Power Circuit

distribution circuit shown in Fig. 2.27. In this circuit inductances $L_A$ and $L_{AB}$, $L_B$ and $L_{BC}$, and $L_C$ and $L_{CA}$ have unity coefficient of coupling, respectively.

## Solution

The netlist solution of this example in Fig. 2.28 shows how to deal with two PSpice problems. The first problem occurs because inductances $L_{AB}$, $L_{BC}$, and $L_{CA}$ form a loop. PSpice does not solve circuit problems that have a loop consisting of voltage sources or inductances. To eliminate this problem, the netlist inserts resistances $R_{ABX} = R_{BCX} = R_{CAX} = 10$ m$\Omega$ in series with inductances $L_{AB}$, $L_{BC}$, and $L_{CA}$, using additional nodes LA1, LB1, and LC1. Using three extra resistances keeps the three-phase circuit balanced. The values of these resistances are small enough not to have much voltage drop, keeping the line voltages in the simulation the same as for the original circuit. The second problem is that the secondary circuit has no ground node (node 0). Either node LA, LB, or LC can be node 0. But this solution to this problem does not deal with each line in the same way. A better solution eliminates the isolation between the load and source circuit by introducing three equal large-valued resistances $R_{AN}$, $R_{BN}$, and $R_{CN}$ connecting between nodes LA, LB, and LC and the ground node. The netlist in Fig. 2.28 uses this solution.

The size of the magnetizing reactance $\omega L_{AB} = \omega L_{BC} = \omega L_{CA}$ is much greater than the load resistances $R_{AB} = R_{BC} = R_{CA}$, so the delta load on the secondary reflects to the primary in proportion to $L_A/L_{AB} = 4$, so $R_A$, $R_B$, and $R_C$ have 40-$\Omega$ resistances in series to ground. Then, the phasor voltage across $L_A$

| FREQ | V(A1) | VP(A1) | V(B1) | VP(B1) | V(C1) |
|---|---|---|---|---|---|
| 6.000E+01 | 2.342E+02 | 1.483E-02 | 2.342E+02 | -1.200E+02 | 2.342E+02 |
| FREQ | VP(C1) | | | | |
| 6.000E+01 | 1.200E+02 | | | | |
| | | | | | |
| FREQ | V(A1,B1) | VP(A1,B1) | V(B1,C1) | VP(B1,C1) | V(C1,A1) |
| 6.000E+01 | 4.056E+02 | 3.001E+01 | 4.056E+02 | -8.999E+01 | 4.056E+02 |
| FREQ | VP(C1,A1) | | | | |
| 6.000E+01 | 1.500E+02 | | | | |
| | | | | | |
| FREQ | I(RLA) | IP(RLA) | I(RLB) | IP(RLB) | I(RLC) |
| 6.000E+01 | 5.850E+00 | -5.935E-01 | 5.850E+00 | -1.206E+02 | 5.850E+00 |
| FREQ | IP(RLC) | | | | |
| 6.000E+01 | 1.194E+02 | | | | |
| | | | | | |
| FREQ | V(LA) | VP(LA) | V(LB) | VP(LB) | V(LC) |
| 6.000E+01 | 6.753E+01 | -2.999E+01 | 6.753E+01 | -1.500E+02 | 6.753E+01 |
| FREQ | VP(LC) | | | | |
| 6.000E+01 | 9.001E+01 | | | | |
| | | | | | |
| FREQ | V(LA,LB) | VP(LA,LB) | V(LB,LC) | VP(LA,LC) | V(LC,LA) |
| 6.000E+01 | 1.170E+02 | 1.483E-02 | 1.170E+02 | -5.999E+01 | 1.170E+02 |
| FREQ | VP(LC,LA) | | | | |
| 6.000E+01 | 1.200E+02 | | | | |
| | | | | | |
| FREQ | I(RAB) | IP(RAB) | I(RBC) | IP(RBC) | I(RCA) |
| 6.000E+01 | 1.170E+01 | 1.483E-02 | 1.170E+01 | -1.200E+02 | 1.170E+01 |
| FREQ | IP(RCA) | | | | |
| 6.000E+01 | 1.200E+02 | | | | |

FIGURE 2.29 Three-Phase Line and Phase Voltages and Currents

is

$$V_{A1} \approx \frac{40}{40+1} \times 240\angle 0° \approx 234\angle 0° \text{ V}$$

while $V_{B1}$ and $V_{C1}$ have the same value, but lag and lead $V_{A1}$ by 120°, respectively. The voltages across $L_{AB}$, $L_{BC}$, and $L_{CA}$ are approximately one half of the voltages $V_{A1}$, $V_{B1}$, and $V_{C1}$. In addition, the solution in Fig. 2.29 shows the typical $\sqrt{3}$ and 30° magnitude and phase relationship between phase and line voltages.

As a special project, change the netlist using additional E-type sources as wattmeters to measure the average power that the circuit delivers to the three load resistances using the two-wattmeter method.

### 2.5.8 Control of Operating Point

The NODESET and IC commands allow you to help PSpice find an operating point. To use the NODESET command, type

.NODESET V(<Node_1>[, <Node_2>]) = <Value_1> ...

This command sets some nodes or node-voltage differences to initial values. PSpice starts with these values and proceeds to the solution. This possibility is useful if PSpice has trouble finding a solution, and you can help by giving one or more values near the solution. Another situation where this command can be useful is when you wish to guide PSpice to another of more than one possible solution. The values that you give apply for operating point calculations for AC, transients, and first DC value of a sweep, but not thereafter.

The IC command has the form

.IC V(<Node_1>[, <Node_2>]) = <Value_1> ...

and clamps the node voltages or node-voltage differences to the values that you specify. PSpice releases the node values after computation begins. The IC command overrides the NODESET command.

### 2.6 Advanced Probe Techniques

This section describes how to enhance plots that you create with Probe and make measurements using Probe's cursor, macrofunctions, and goal functions.

### 2.6.1 Probe Files

Probe requires a file with a DAT suffix, which PSpice generates when a PROBE statement exists in a netlist. However, there are six other files that Probe also can use. A list of these file types appears in Table 2.17, with descriptions of how Probe uses each. To run Probe on an IBM PC or compatible computer using any of these files, include the option notation after typing PROBE. For example, to run Probe with the file MYCIRCUIT.DAT and have the macros and goal functions from MYCIRCUIT.MAC and MYCIRCUIT.GF be available, after the DOS prompt within the PSpice/Probe subdirectory, type

| Option | Suffix | Default Name | File Function |
|--------|--------|--------------|---------------|
| /C | CMD | | A command file to run Probe |
| /D | DEV | PSPICE.DEV | Defines display and hard-copy devices |
| /G | GF | PROBE.GF | Defines goal functions for performance analysis |
| /L | LOG | | Records actions that you can replay |
| /M | MAC | PROBE.MAC | Defines macro functions |
| /S | DSP | PROBE.DSP | Allows display of previous sessions |

Table 2.17 Probe Files

PROBE /M MYCIRCUIT.MAC /G MYCIRCUIT.GF MYCIRCUIT.DAT

You can omit the suffix for any of these files unless another file having the same name without the suffix exists. After the option indicator, you can omit the file name and Probe then uses the default name for that file type. For example, with no file name after /M, Probe uses PROBE.MAC instead of MYCIRCUIT.MAC. The order of the file options does not matter, but the DAT file name must appear last.

To run Probe on a Macintosh with any of these auxiliary files, give each auxiliary file the same name as your DAT file and include the appropriate suffix. Place these auxiliary files within the same folder as the DAT file. For example, if your DAT file is MyCircuit.DAT and its folder has a file MyCircuit.GF, then Probe includes MyCircuit.GF when you open MyCircuit.DAT from Probe.

### 2.6.2 Startup and Section Selection Menus

This section describes the sequence of menus that you encounter as Probe begins to run. To select items from these menus, you can use any of the following methods:

- Click on the item with the mouse

- Type the letter that appears in capitalized form in the item's name (e.g., E for Exit_program)

- Use the left <←> or right <→> arrow keys to highlight the selection, then type the <Enter> or <Return> key

- With a Macintosh, select an item from the Probe Menu on the menubar

Typing the <Escape> key returns you to the previous menu. The <Escape> key also provides a way to avoid typing an entry, when your selection requires a response that you do not want to make.

After you choose a DAT file for Probe to use, you may encounter the following sequence of menus :

- Startup

- Section Selection

- Analog/Digital

The Startup menu appears only if your netlist contains two or more DC, AC, or TRAN commands. The four options available on this menu are

- Exit_program

- Dc_sweep

- Ac_sweep

- Transient_analysis

Probe allows plotting of only one procedure (DC, AC, or TRAN) at a time, so you must choose which you want to use at this time. If your netlist has only one type of analysis, then Probe bypasses this menu.

If your netlist contains multiple sections for the analysis procedure that you select from the Startup menu, you next encounter the Section Selection menu. Here you have three choices:

- Exit

- all_<Analysis>

- Select_sections

Depending on the analysis type, **all_<Analysis>** is either **all_Dc_sweep**, **all_AC_sweep**, or **all_Transient_analysis**. If you choose **Select_sections**, the item **Select_all_sections** in the section-selection box becomes highlighted. Use the up ($\uparrow$) and down ($\downarrow$) arrow keys to change the highlighting from one section to the next. Type the spacebar to select a highlighted item. Repeat this process to make multiple selections. To cancel selection of an item, highlight the item and type the spacebar. To select all but a few items, choose **Select_all_sections** and remove the few you do not want. If you have more sections than appear in the selection box, a plus (+) sign appears next to the last item in the box. To see more sections when the selection box is full, highlight the last item and type the down-arrow key to make the list of items advance about half of a screen. When your selections are complete, either click the **Exit** item or type E to proceed to the Analog/Digital menu. If you select **all_<Analysis>**, operation goes to the Analog/Digital menu with all sections available for plotting. Probe bypasses the Section Selection menu when your analysis has only one section.

Probe now shows the Analog/Digital menu. The Analog/Digital menu has some or all of the following items:

- Exit

- Add_trace

- Remove_trace

- X_axis

- Y_axis

- Plot_control

- Display_control

- File

- Macros

- Hard_copy

- Cursor

- Zoom

- Label

- Goal_functions

Chapter 1, Sec. 1.4 describes the Analog/Digital menu briefly and shows how to use **Add_trace**. The following sections describe other features of the Analog/Digital menu that you might not discover by yourself.


### 2.6.3 Axis  Control

The X_axis and Y_axis menus contain the following common selections:

- Exit

- Log/Linear

- Auto_range

- Set_range

The **X_axis** and **Y_axis** menu selections on Probe's Analog/Digital menu allow you to choose to have linear or log plots by selecting the **Log** or **Linear** item. If you have a linear plot, this item appears as **Log**. After selecting **Log**, it changes to **Linear**. For example, from the X_axis menu typing character L causes the x-axis to change from log to linear if in log mode, and vice versa.

To set the range of the variable use **Set_range**. If the range has been set, **Auto-range** becomes a menu option. Using **Set_range** allows you to set the range of the x- or y-axis variable. After choosing this command enter the range by typing two numerical values and <Enter> or <Return>. Separate the values by one or more blanks, or by a comma. Enter the values using floating point notation, exponential notation, or use a suffix scale factor (f, p, n, u, m, k, M, T). Remember that Probe uses m for milli and M for Mega, not  M and MEG as in PSpice. Leave no blanks between the number, the suffix, or the unit. If the first number exceeds the second, then the axis reverses. Choose **Exit** to return to the Analog/Digital menu.

The X_axis menu has the following items, not on the Y_axis menu:

- Restrict_data/Unrestrict_data

- X_variable

- Fourier/Quit_fourier

- Performance_analysis/Quit_performance_analysis

- Options

    - Exit

    - Histogram_divisions

    - Show_histogram_statistics/Do_not_show_histogram_statistics

Changing the x-axis variable using **X_variable** may provide a significant plot. For example, with a nested DC sweep, the first sweep variable is the x-axis variable; and the second sweep variable is each curve's parameter. You can change the x-axis variable to the second sweep variable if you wish. With transient analysis, time initially is the x-axis variable. In AC analysis you can change the x-axis variable to be another voltage or current to see how the y-axis variable relates to the x-axis variable as the frequency changes. Using **Restrict_data** sets the range of x-axis data for Fourier analysis or the range functions s(x), avg(x), rms(x), min(x), or max(x). **Unrestrict_data** removes a restriction set by **Restrict_data** and appears only after a restriction is in place. Sections 2.6.8, 2.6.9, and 2.6.10 describe how to use Performance_analysis, Fourier, and Histogram items.

The following items occur on the Y_axis menu, but not on the X_axis menu

- add_aXis
- Remove_axis
- Change_title
- selecT_axis
- color_Option
    - Exit
    - Normal
    - Match_axis
    - Sequential_per_axis

On the **Y_axis** menu you can choose **add_aXis** to insert up to two more vertical axes for each plot. Each new axis becomes selected, as shown by the characters "SEL>>" along the new axis. When you have more than one axis **Remove_axis** and **selecT_axis** appear among the menu items of the Y_axis menu. **Remove_axis** removes the selected axis. **selecT_axis** lets you select an axis using the right <→> or left <←> arrow keys. However, it is easier to select the axis with the mouse by clicking the (left) mouse button when the cursor is just to the left of the desired axis, which can be done with the Analog/Digital menu showing.

### 2.6.4 Plot Control

Selecting **Plot_control** on the Analog/Digital menu opens the Plot Control menu. This menu contains the following items:

- Exit
- Add_plot
- Remove_plot
- Select_plot
- unsYnch_plot
- always_Use_symbols/Never_use_symbols
- auTo_symbols/Never_use_symbols

• Mark_data_points/Do_not_mark_data_points

If you select **Add_plot**, a new plot appears in Probe's window. Keep adding as many plots as you wish. Eventually Probe decides that the plots are too small and **Add_Plot** disappears from the available menu items. Two or three plots are probably enough. When two or more plots coexist in Probe's window, **Remove_plot** becomes available. Before selecting this option, be sure that the plot you want to remove is the active plot. When you have two or more plots, **Select_plot** lets you select the plot that is active using the up arrow <↑> and down arrow <↓> keys. Unlike the right arrow <→> or left arrow <←> key selection of axes, which cycle, these do not cycle past the top or bottom plots. However, selection of a plot by clicking the mouse with the cursor within the plot area is easier and can be done anytime that you wish.

Selecting **unsYnch_plot** gives a plot its own x-axis. When a selected plot already is "unsynched," the **unsYnch_plot** command disappears. To "resynch" a plot, remove the plot with **Remove_plot** and use **Add_plot** and **Add_trace** to create a "synched" plot.

The items **always_Use_symbols** and **Never_use_symbols** toggle with **autO_symbols**. Select **always_Use_symbols** to force marking plots with symbols. Use **Never_use_symbols** to remove identification symbols from curves. PSpice starts with Item **autO_symbols**, which marks curves if there are more traces than colors available for drawing and less than nine traces. **Mark_data_points** and **Do_not_mark_data_points** also toggle, adding and removing the actual PSpice data points. You probably do not need to see the data points when there is much data. When there are few data points, displaying them shows that the curve breaks, not because of some mysterious circuit effect but because of the sparse number of points that Probe connects with straight lines.

### 2.6.5 Cursor Control

With one or more traces on a plot, selecting the Cursor menu item of the Analog/Digital menu activates two cursors. Initially, both cursors mark the start of the first trace. Use the right <→> or left <←> arrow key to move the first cursor. Typing either arrow key moves the cursor one pixel. Holding either arrow key causes the cursor to jump continuously by 10-pixel increments. Holding down the <Shift> key or setting the <Caps Lock> key and typing or holding the right <→> or left <←> arrow key moves the second cursor. Typing the <Home> key or the <End> key moves the first cursor to the beginning or end of the trace. Holding down the <Shift> key, or setting the <Caps Lock> key, while typing <Home> or <End> does the same for the second cursor. You also can move the first cursor by clicking or dragging with the mouse. Clicking the mouse button with the mouse cursor away from the vertical cursor line makes the cursor jump to the mouse-cursor position. Clicking and dragging pulls the cursor along with the mouse cursor. Again, use the <Shift> or <Caps Lock> key to move the second cursor with the mouse.

To detach the first cursor from its current trace and attach it to an adjacent trace, PC users hold the <Control> key down while typing the right <→> or left <←> arrow key. Macintosh users hold the <Option> key instead of the <Control> key. This action cycles the cursor attachment to the last or first trace and then skips around to the first or last trace. To change the second cursor's trace attachment, hold the <Shift> key or set the <Caps Lock> key while using the right <→> or left <←> arrow keys. As the cursors move, their x- and y-coordinate values and the differences of these between cursors appear in a box at the lower right of the Probe window. You also can attach the first cursor to a new trace by clicking on the identification symbol of the new trace in the trace label below the horizontal axis. To attach the second trace this way, hold down the <Shift> key or set the <Caps Lock> key.

The Cursor menu has the following items:

- Exit

- Hard_copy

- Peak

- Trough

- slOpe

- Min

- maX

- Search_commands

- Label_point

The menu items **Min** and **maX** move the cursors to the trace horizontal position having the minimum and maximum y-axis value, respectively. The menu items **Peak**, **Trough**, and **slOpe** are directionally sensitive. Their action depends on the current direction of the cursor. After first selecting **Cursor** from the Analog/Digital menu, both cursors expect to move forward. Thereafter, their last direction of motion is their current direction. Menu item **Peak** moves the cursor in the current cursor direction to the next point on the current trace having two adjacent points with lower y-axis values. Menu item **Trough** moves the cursor in the current cursor direction to the next point on the current trace having two adjacent points with higher y-axis values. Menu item **slOpe** moves the cursor in the current cursor direction to the next point on the current trace having the largest positive or negative slope value. Choose these actions to apply to the second cursor using either the <Shift> key or the <Caps Lock> key. Selecting **Label_point** marks the plot with the horizontal and vertical coordinates of the cursor that you choose using the <Shift> or <Caps Lock> key. You can manipulate these labels after returning to the Analog/Digital menu by selecting the **Label** item.

The **Search_commands** item requires entry of a text search command that tells Probe where to place the cursor that you select. Holding down <Shift> or setting the <Caps Lock> key before selecting the **Search_commands** item makes the second cursor respond to the search command that you type. Also, be sure that the cursor you select is attached to the waveform you want to measure. The syntax of a search command is

Search [<direction>] [/<Start_Point>/] [#<Consecutive_Points>#]
      [(<Range_x>)[,(<Range_y>)]] [for] [<Repeat>:] <Condition>]

As usual, square brackets [] denote optional items. Replace all items within <...> including the inequality brackets < > themselves with appropriate text or values.

The keyword "Search" can be abbreviated to an upper- or lowercase character S. Replace <direction> with "Forward" or "Backward," which abbreviate as "f" or "b." Forward means to search in the direction of increasing value of the independent variable, which may be backward if you have the x-axis variable increasing to the left. Replace <Start_Point> with one of the following;

| | |
|---|---|
| ^ | First point within search range |
| Begin | First point within search range |
| $ | Last point within search range |
| End | Last point within search range |
| xn | Marked point (e.g., x2) – Only with goal functions in Performance_analysis |

By default, the current point is the starting point. <Consecutive_Points> is the number of points in sequence for a condition to be met. <Range_x> and <Range_y> limit the x- and y-range of the search. The values can be in floating-point notation, a percent of the full-range value, or marked-point notation when doing a Performance_analysis. By default, the range includes all visible points in the plot. <Repeat> specifies which numerical occurrence of <Condition> to find. If the value of <Repeat> exceeds the numerical occurrences of <Condition>, then the search command finds the last <Condition>. <Condition> must be one of the following:

LEvel (<Value>[,<Pos_Neg>]) finds the first point in the search direction having at least <Consecutive_Points> -1 beyond the level.

> <Value> can be as follows:
> A floating-point number (e.g., 100k, 1e5, ...)
> A percent (e.g., 35%)
> A marked point [e.g., y2 (see Sec. 2.6.6, Performance Analysis)]
> A value relative to max or min (e.g., max - 5)
> A dB value relative to max or min (e.g., max - 3 db)
> A relative value (e.g., . -5 is the last value minus 5)
> A dB relative value (e.g., . -3 dB is 3 dB less than the last value)

> <Pos_Neg> can either be Positive (P) or Negative (N) or Both (B), which is the default.

SLope [<Pos_Neg>] finds the next maximum slope, either positive or negative. Here, the default of <Pos_Neg> is positive.

PEak finds the nearest maximum in the search direction having #<Consecutive_Points># on each side. Type the # symbol to bracket your entry of the number of consecutive points.
TRough finds the nearest minimum in the search direction having #<Consecutive_Points># on each side.

MAx finds the maximum y-axis value within <Range_x>. If two equal-valued maxima exist within the range, this condition finds the first in the search direction. The options <direction>, #<Consecutive_Points>#, or <Repeat> have no effect.

MIn finds the minimum y-axis value within <Range_x>. If two equal-valued minima exist within the range this condition finds the first in the search direction. The options <direction>, #<Consecutive_Points>#, or <Repeat> have no effect.

XVal (<Value>) where <Value> can be as above. Neither <direction>, <Consecutive_Points>, <Range_x>, <Range_y>, or <Repeat> affect this command.

You can abbreviate each condition using the two capital letters, e.g., PE is the abbreviation of PEak.

## Example 2.10 Search Command Analysis of a Resonant Circuit

Use PSpice to simulate the circuit shown in Fig. 2.30 when resistance $R_1$ varies from 0.5 k$\Omega$ to 2.0 k$\Omega$ in 0.1-k$\Omega$ steps. With Probe plot the inductance current $i_{L1}(t)$ when $R_1$ equals 1.5 k$\Omega$. With the cursor at the x-axis origin, use **Search_commands** from the Cursor menu to find the third occurrence of 0.95 mA and the second peak value.

### Solution

A PSpice netlist to simulate the circuit appears in Fig. 2.31. To select <t_Print> and <t_Final> find the natural frequencies from the characteristic equation

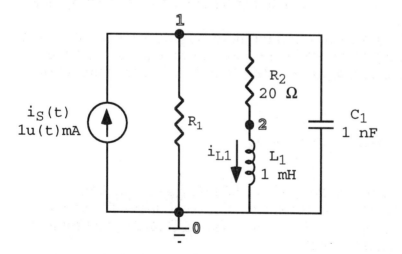

FIGURE 2.30 Resonant Circuit

```
Example 2.10 - Search Command Analysis of a Resonant Circuit
*
.OPT   NOBIAS
.PARAM        Rval=0.5
*
Is      0      1      1m
R1      1      0      {Rval}
R2      1      2      20
L1      2      0      1m
C1      1      0      1n
.TRAN 100n   20u    0         100n   UIC
.STEP PARAM Rval    0.5k   2.0k   0.1k
.PROBE
.END
```

FIGURE 2.31 PSpice Netlist for Example 2.10

$$s^2 + \left(\frac{1}{R_1 C_1} + \frac{R_2}{L_2}\right)s + \left(1 + \frac{R_2}{R_1}\right)\frac{1}{L_1 C_1} = s^2 + 2\alpha s + \omega_o^2 = 0$$

Then

$$\alpha = \frac{1}{2}\left(\frac{1}{R_1 C_1} + \frac{R_2}{L_1}\right) \approx 0.3\,\mu s$$

The circuit response damps out in about 15 μs, so we select <t_Final> equal to 20 μs. The damped natural frequency is

$$\omega_n = \sqrt{\left(1 + \frac{R_2}{R_1}\right)\frac{1}{L_1 C_1} - \alpha^2} \approx 1\,\text{Mrad}/s$$

which corresponds to a period of approximately 6 μs. Choosing <t_Print> equal to 100 ns gives about 50 or 60 points per period, which should display the oscillations effectively.

Choosing to plot only the waveform for the case where $R_1$ equals 1.5 kΩ and using Probe gives the plot shown in Fig. 2.32. Select **Cursor** from the Analog/Digital menu and **Search_commands** from the Cursor menu. Now, enter

Search forward 3: Level(0.95m)

which you can abbreviate and concatenate as

sf3:le(0.95m)

to locate the third occurrence of $i_L(t)$ equal to 0.95 mA. Select the item **Label_point** on the Cursor menu to label this point. Next, move the cursor back to the x-axis origin using the <Home> key. Select **Search_commands** from the Cursor menu. This time enter

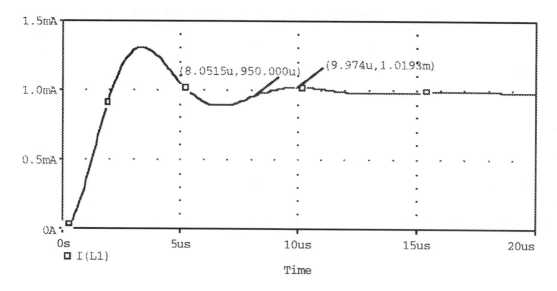

FIGURE 2.32 Probe Plot of $i_L(t)$

Search forward 2: Peak

or

sf2:pe

to find the second peak value of $i_L(t)$ from the origin. Label this point using the **Label_point** item on the Cursor menu. Return to the Analog/Digital menu and use the Label item and **Move** to move the point labels so that they do not overlap. Explore all of the cursor control features and use the search command syntax to make sure that you understand how the search-command syntax works.

## 2.6.6 Functions

Selection of **Add_trace** on Probe's Analog/Digital menu lets you specify either variables or functions to plot. Probe uses the same variable notation as the PRINT and PLOT commands of PSpice, except that you cannot refer to the voltage across a device. For example, if a resistance $R_5$ connects between nodes 2 and 8, use V(2,8), not V(R5) in Probe. When more than one set of data is available in Probe you can select a specific set using the @ <Set_No> notation. For example, if your netlist has a STEP statement with three values of a parameter, you can refer to the voltage at node 5 in the second step by writing

V(5)@2

You can use the name of a sweep variable in a function. With AC analysis, the sweep variable is FREQUENCY. In transient analysis, TIME is the sweep variable. In addition to using the arithmetic operators +, -, *, and / in functions, you can use parentheses to control operator priority. Probe has all of the

| Function | Expression | Comment |
|---|---|---|
| sgn(x) | 1 x>0, -1 x<0, 0 x=0 | |
| m(x) | Magnitude of x | x may be complex |
| p(x) | Phase of x | In degrees |
| r(x) | Real part of x | |
| img(x) | Imaginary part of x | |
| g(x) | Group delay of x | |
| d(x) | Derivative of x | With respect to X_axis |
| s(x) | Integral of x | With respect to X_axis |
| avg(x) | Running average of x | With respect to X_axis |
| avgx(x, d) | Running average of x from x-d to x | Over range of X_axis |
| rms(x) | Running RMS average of x | Over range of X_axis |
| db(x) | Magnitude of x in decibels | |
| min(x) | Minimum of real part of x | |
| max(x) | Maximum of real part of x | |

Table 2.18 Additional Probe Functions


functions of PSpice (see Table 2.8) and also the additional functions shown in Table 2.18.

Probe does complex arithmetic and displays the magnitude of the result. For example, typing

V(2)+V(5)

gives a trace that displays all of the points using

$|[Re\{V(2)\}+Re\{V(5)\}] + j[Im\{V(2)\}+Im\{V(5)\}]|$

and typing

V(2)*V(5)

produces a trace that displays all of the points using

|V(2)|*|V(5)|

On the other hand

DB(V(1,2)+V(2))

is not the same as

VDB(1,2)+VDB(2)

You can use an entire trace as an argument of a subsequent function using the #<Trace_No> notation. For example, if you want to double the second trace of a set of five traces already in a plot, type

2*#2

Typing

#2*#3

gives the product of each data element in set 2 times the corresponding element in data set 3.

### 2.6.7 Macros

With macros you can write functions using the arithmetic operations +, -, *, and /, parentheses, and any of the functions from Tables 2.8 and 2.18 to create new functions. You can create these macros "on-the-fly" by selecting **Macros** on the Analog/Digital menu. Probe saves these macros in the current MAC file. Also, you can create macros using a word processor to edit a MAC file that you load when you run Probe. See Section 2.6.1 for details on how to have Probe load your macro file. The syntax for a macro is

<Macro_Name>[(<Arg_1>,...)] = <Definition>

Your choice for <Macro_Name> can be any alphanumeric name that does not conflict with function names in Tables 2.8 or 2.18. Probe even lets you begin a macro name with a numeral. With one or more arguments after <Macro_Name>, enter the arguments within parentheses, type the equality sign (=) and write the macro <Definition> using the arithmetic operators, parentheses, Probe functions, and other macros. Probe does not allow recursive macro definitions. Blank spaces improve readability, but macros have a one-line, 80-character limit, so use of blanks may be unwise. Blank lines have no significance in a MAC file. A line beginning with an asterisk (*) is a comment and has no effect. Remember to include an asterisk at the beginning of the title line in your macro file. In-line comments that a semicolon demarks are useful. Some examples of useful macros are

pi          = 4*atan(1)

twopi       = 2*pi; Provided pi exists, otherwise use 8*atan(1)!

sinh(x)     = 0.5*(exp(x) - exp(-x))

cosh(x)     = 0.5*(exp(x) + exp(-x))

t           = time; Simplifies entry

tm          = 1000*time; Allows entering equations with t in ms.
                  or for AC simulations

w           = twopi*frequency; Assuming twopi exists.

conj(z)     = m(z)*m(z)/z; Computes the complex conjugate of z.

avgpwr(v,i) = r(v*conj(i); Assuming conj exists.

### 2.6.8 Performance Analysis and Goal Functions

Probe's **Performance_analysis** item on the X_axis menu makes possible plots of selected waveform evaluations as a function of a changing parameter. This feature makes the STEP, TEMP, and MC commands very significant features of PSpice. Performance analysis uses goal functions, which you also can apply to single traces with the **Goal_functions** item on the Analog/Digital menu. To do performance analysis, with multiple sets of data available, select

**X_axis** from the Analog/Digital menu and then **Performance_analysis** on the X_axis menu. Probe then creates a plot that has the parameter that differs from one set to the next as the independent variable. Next, select **Add_trace** from the Analog/Digital menu and enter the name of a goal function using expression arguments and substitution arguments appropriate to the goal function. The goal function has to exist in a GF text file and be available for the current Probe session.

　　　To create a goal function, use your word processor to create a text file that describes all of the goal functions that you plan to use with a Probe session. Give your goal function file the name PROBE.GF or use the same name as your DAT file with the GF suffix (see Sec. 2.6.1).[3] A goal function file can have blank lines and comments, using an asterisk (*) in the first column of the comment line. The syntax of a goal function is

```
<Goal_Function_Name>(1,[2,…,n,<Sub_Arg1>,…]) =
        <Marked_Point_Expression>
{
  1|
    <Search_Commands_and_Marked_Points_for_1>
  ;
  [2|
    <Search_Commands_and_Marked_Points_for_2>
  ;
  etc.]
}
```

Lines within a goal function definition may continue to the next line without any need to insert a special character. Line spacing and indentation in the syntax description above is useful but unnecessary. However, do not use tabs for indentation. Replace <Goal_Function_Name> with a name of your choice using alphanumeric characters (A–Z and 0–9) and the underscore (_), but with less than 50 characters and beginning with an alphabetic character. The name is not case sensitive. The expression arguments 1,2,…,n identify how many Probe expressions to give when you call the goal function. For example, if your GF file has the goal function "period(1)" then

　　　period(V(4)/I(L1))

uses the Probe expression V(4)/I(L1) as the first and only argument of goal function "period." The expression argument(s) can use +, -, *, /, parentheses, any Probe variable names, and all of the Probe functions from Tables 2.8 and 2.18 that return an array of values. The expression arguments that you supply represent numerical arrays that you might plot with Probe. If you wish to use any substitution arguments (<Sub_Arg1>,…), use unique names that follow the same convention as for goal functions. These argument names also appear as arguments in the search commands in the body of your goal function definition.

---

[3] MicroSim's documentation suggests that Macintosh users can let GF files have the name Probe.GF. This does not work with Version 6.0. Probe's **File** menu item **Change Goal Function File …** allows you to select a goal function file after loading a DAT file. But the current DAT file cannot access these goal functions. After changing the GF file, close the current DAT file and open it again. Now the goal functions in the GF file are available.

When you call the goal function you provide appropriate values in the position of each substitution argument. The function uses these values in place of the argument name within the function. In the body of the function and within the range of nl and the next semicolon (;) the nth Probe expression serves as the operand of the various search commands. When a search command in the goal function identifies a point P1 having coordinates x1, y1, this point can be marked using the notation !1. For example, in the goal function "period" the following lines identify two points $P_1$ and $P_2$ with coordinates $x_1$, $y_1$ and $x_2$, $y_2$

```
1|
  search forward /b/ 2:level(0,p) !1
  search forward 2:level(0,n) !2
;
```

which identify the first positive-going and negative-going zero crossings of whatever expression passes into the goal function "period" as the expression argument. Notice that there is no necessary correlation of expression numbers with marked point numbers. The <Marked_Point_Expression> has a single numerical real value that relates to a single trace in the STEP, TEMP, or MC set of traces. A marked point expression is the same as a normal Probe expression, except that it uses marked-point coordinates instead of circuit variables. The arithmetic operators +, -, *, and / all apply and parentheses override the normal hierarchy of operations. All functions from Tables 2.8 and 2.18 are usable, if they return a scalar value for a scalar argument. In addition there is a function $mpavg(x_b, x_e, f)$ which finds the average of all points between marked points having beginning x-value $x_b$ and ending x-value $x_e$. Parameter f changes the range $x_b$, $x_e$ by factor f. Specifically, mpavg averages all points within

$$\frac{x_b + x_e}{2} \pm \frac{f}{2}\left(x_e - x_b\right)$$

Figure 2.33 shows two examples of goal functions. Goal function "risetime" has one expression argument. The search functions find and mark the expression value at 0.1 mA and 0.9 mA, which are approximately the 10% and 90% values for inductance current $i_{L1}(t)$ of Example 2.10. The marked-point expression then takes the difference between the larger and the smaller x-axis values. You can generalize this function by using two substitution arguments in the defining line

    risetime(1,val1,val2) = x2-x1

and replacing 0.1m with val1 and 0.9m with val2. Then, when you use the general form, supply the numerical values for val1 and val2. For example, if expression is V(5)/V(1), which changes from 1 to 5, then when you use "risetime" write

    risetime(V(5)/V(1), 1.4, 4.6)

The second goal function "overshoot" finds and marks the maximum and the next trough and uses function mpavg to average points close to the midpoint between these points. This value approximates the steady-state value. The marked-point expression then uses this value and the maximum value $y_1$ to evaluate the overshoot percent.

```
* Goal Function Definitions for Example 2.11
risetime(1)=x2-x1
{
  1|
    search forward for level (0.1m) !1
    search forward for level (0.9m) !2
  ;
}
overshoot(1)=100*(y1/mpavg(x1,x2,0.1)-1)
{
  1|
    search for max !1
    search for trough !2
  ;
}
```

FIGURE 2.33 Goal Functions for Example 2.11

## Example 2.11 Using Goal Functions

a) Use the DAT file from Example 2.10 with Probe to plot the waveforms for $i_L(t)$ for the cases where $R_1$ equals 0.5, 1.0, 1.5, and 2.0 k$\Omega$.

b) Including all cases where there is an overshoot, plot the risetime and the overshoot of $i_L(t)$ as a function of $R_1$. Use the risetime and overshoot goal function definitions shown in Fig. 2.33.

### Solution

a) The Probe plot of $i_L(t)$ for the four cases appears in Fig. 2.34. Create this plot by selecting only the four specific cases from the Section Selection menu that appears after choosing the file EX_2_10.DAT either at startup of Probe with a PC or using the File menu with a Macintosh after starting Probe. Alternately,

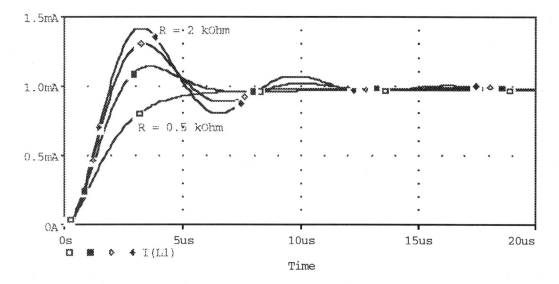

FIGURE 2.34 Plot of $i_{L1}(t)$ for $R_1$ Equal to 0.5, 1.0, 1.5, and 2.0 k$\Omega$

load all sections and then use **Add_trace** four times, specifying in sequence I(L1)@1, I(L1)@6, I(L1)@11, and I(L1)@16. Use the **Label** item on the Analog/Digital menu to mark the first and last curves with "R = 0.5 kOhm" and "R = 2 kOhm." Notice that when $R_1$ equals 0.5 kΩ there is no overshoot. Plotting the first four cases with $R_1$ ranging from 0.5 kΩ to 0.8 kΩ shows that the 0.6-kΩ case does have an overshoot.

b) Returning to Probe's starting menu, select **Transient_analysis**. Then choose **Select_sections**. Since we want all but the first section, save time by choosing **Select_all_sections** and then select the section for Rval equal to 500 Ω, which removes the selection of the 500-Ω Rval section. Choose **Exit** to move to the Analog/Digital menu. Now, select **X_axis** on the Analog/Digital menu and on the X-Axis menu select **Performance_analysis**. These actions return you to the Analog/Digital menu, but the plot's horizontal axis now is Rval. Choose **Add_trace** and enter

> risetime(I(L1))

to obtain a plot of the risetime as a function of the resistance Rval. Next, use the Y_axis menu to create a new y-axis. Keeping this new axis selected, choose **Add_trace** and enter

> overshoot(I(L1))

to see the overshoot as a function of resistance Rval. Label the axes and plots to see the curves as in Fig. 2.35. The curves show that the risetime falls from about 2.77 μs to about 1.28 μs while the overshoot increases from approximately 0.8% to about 33.5% as resistance $R_1$ increases from 600 Ω to 2.0 kΩ. Deriving these results from the circuit's equations would be a very difficult task!

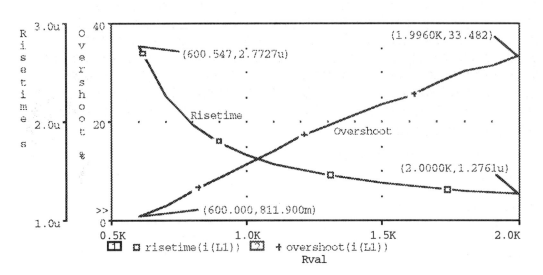

FIGURE 2.35 Risetime and Overshoot for Example 2.11 Using Goal Functions and Performance Analysis

### 2.6.9 Fourier   Transform

When displaying a transient as a function of time in Probe the X_axis menu has a **Fourier** item. Selecting this item converts the time axis to a frequency axis. The dependent variable becomes a numerical approximation of the Fourier spectrum of the original variable. The Fourier transform has a real and imaginary part, so you can plot either the real, imaginary, magnitude, or phase of the transform using the usual Probe notation. The spectrum has values at multiples of the resolution, which is a frequency value equal to the reciprocal of the time duration <t_Final> of the transient simulation. The extent of the transform is the highest frequency value that Probe computes for the spectrum. The extent depends on the number of points in the PSpice simulation as well as on <t_Final>. You can affect the number of points by setting the maximum step size <t_Ceil>, but the number of points really is out of your control. Probe rounds the number of points up to the next higher power of 2 and generates a uniformly spaced set of data with this many points for the duration <t_Final>. The extent equals the product of this number of points and the resolution divided by 2. The spectrum shows the amplitudes at each frequency and connects adjacent points with straight lines. When the transient is periodic, the values at each multiple of the resolution frequency are the Fourier series coefficient values (real, imaginary, magnitude, or phase).

### 2.6.10 Histograms

Probe displays histograms (Version 5.3 and higher) for data resulting from performance analysis of PSpice Monte Carlo simulations. Histograms show the number or percent of occurrences of each value of a variable resulting from variations of the circuit parameters during the Monte Carlo run. To show a histogram, the PSpice netlist has to contain a Monte Carlo simulation using the MC command. In addition include a DC, AC, or TRAN command. To see the histogram display with Probe, include all sections with the Section Selection menu. Then, go to the X_axis menu from the Analog/Digital menu. Now, select **Performance_analysis**. Back at the Analog/Digital menu now, use **Add_trace** to enter a goal function, which exists in an appropriate goal function file. Select the **Options** item from the X_axis menu and choose **Histogram_divisions** to set a specific number of divisions for the histogram display. By default, the histogram contains ten divisions. **Do_not_show_- histogram_statistics** and **Show_histogram_statistics** toggle and allow you to include or exclude these statistics. The statistics that appear include

- Number of samples
- Number of histogram divisions
- Minimum
- Maximum
- Mean
- Median

- 10th percentile
- 90th percentile
- Standard deviation

## Example 2.12 Monte Carlo Analysis and Histograms

Generate a ten-division histogram for the circuit shown in Fig. 2.36 to show the statistical variation of the voltage at node 3 when the resistances have 10% device (DEV) tolerance and with (a) a uniform distribution;  b) a Gaussian distribution; and (c) the BIMODAL distribution of Sec. 2.5.5.

### Solution

To achieve the goal of the exercise, sweep the voltage $V_{CC}$ from 10 V to 12 V using a 2-V step. A DC sweep is necessary so that Probe has more than one DC section to plot. The netlist shown in Fig. 2.37 uses a resistance model to set the device tolerance to 10%.

Run the netlist EX_2_12.CIR using PSpice to create the file EX_2_12.DAT. With your word processor, write the goal function

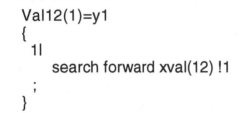

```
Val12(1)=y1
{
  1|
      search forward xval(12) !1
  ;
}
```

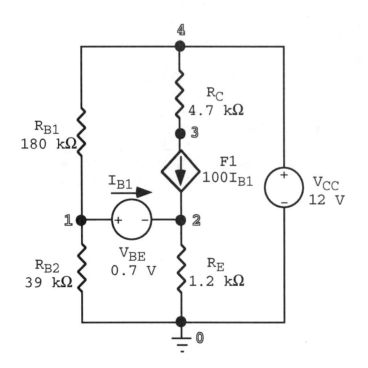

Fig. 2.36 Circuit for Example 2.12

```
Example 2.12 - Monte Carlo Analysis and Histograms
*
* a) Uniform distribution by default
* b) Gaussian distribution
*      .OPTIONS DISTRIBUTION GAUSS
* c) Bimodal distribution
*      .DISTRIBUTION BIMODAL (-1,1) (0,0) (1,1)
*      .OPTIONS DISTRIBUTION BIMODAL
*
VCC      4      0      12
VBE      1      2      0.7
F1       3      2      VBE    100
RB1      4      1      Rmod   180k
RB2      1      0      Rmod   39k
RC       4      3      Rmod   4.7k
RE       2      0      Rmod   1.2k
*
.MODEL        Rmod   RES    (R=1 DEV 10%)
*
.MC    50     DC     V(3)   YMAX   OUTPUT ALL
.DC    VCC    10     14     2
.PRINT DC     V(3)   I(RC)
.PROBE V(3)
*
.END
```

Fig. 2.37 Netlist for Example 2.12

This goal function measures the response value when the independent variable $V_{CC}$ is 12 V, using the Probe expression that replaces the argument of Val12. Save the goal function file as a text file using the name EX_2_12.GF.

Now, run Probe with EX_2_12.DAT as the input file, making EX_2_12.GF available. From the Section Selection menu, choose **All_Dc_sweep**. When you see the Analog/Digital menu, select the **X_axis** item and choose

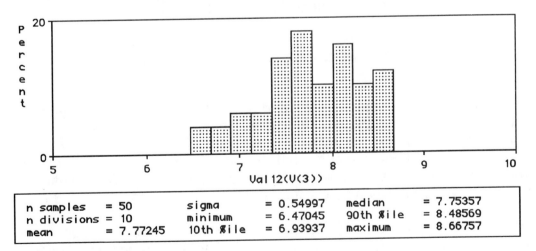

| n samples | = 50 | sigma | = 0.54997 | median | = 7.75357 |
| n divisions | = 10 | minimum | = 6.47045 | 90th %ile | = 8.48569 |
| mean | = 7.77245 | 10th %ile | = 6.93937 | maximum | = 8.66757 |

a) Uniform Distribution.

FIGURE 2.38 Histogram of V(3) Values

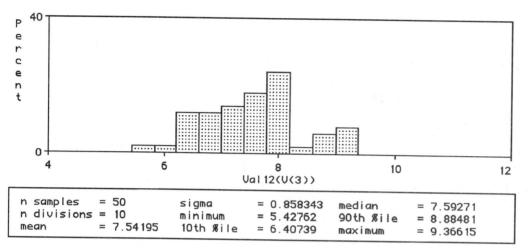

| n samples = 50 | sigma = 0.858343 | median = 7.59271 |
| n divisions = 10 | minimum = 5.42762 | 90th %ile = 8.88481 |
| mean = 7.54195 | 10th %ile = 6.40739 | maximum = 9.36615 |

b) Gaussian Distribution.

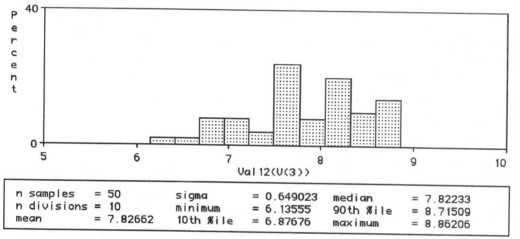

| n samples = 50 | sigma = 0.649023 | median = 7.82233 |
| n divisions = 10 | minimum = 6.13555 | 90th %ile = 8.71509 |
| mean = 7.82662 | 10th %ile = 6.87676 | maximum = 8.86206 |

c) Bimodal Distribution.

FIGURE 2.38 Histogram of V(3) Values (Continued)

**Performance_analysis** from the X_axis menu. Now, back in the Analog/Digital menu, select **Add_trace** and enter

   Val12(V(3))

either by typing directly or typing <Help> and selecting Val12( ) from the menu of available Probe variables and goal functions. This process creates Fig. 2.38a. For Part b, repeat the entire process, removing the asterisk for the Part b Gaussian OPTIONS command. For Part c, restore the asterisk for the Part b statement and remove the asterisks for both Part c commands. The resulting histograms for the Gaussian and the bimodal distribution appear in Fig 2.38b and c.

   The histograms shown in Fig 2.38a, b, and c show how the voltage at node 3 may vary when the resistances have 10% device tolerance with uniform, Gaussian, or bimodal distributions. The output file shows that the value of V(3) is 7.593 V when all resistance values are nominal.

# Chapter 3 PSpice Problems

This chapter provides PSpice problems for you to work. Most of these problems relate to examples, exercises, or problems in *Circuit Analysis* by Cunningham and Stuller. Completing these PSpice problems will help to give you insight into circuit behavior and will develop your understanding of PSpice. Simulation insight complements analytic skills that you obtain from finding circuit solutions using analysis techniques and a calculator. Since PSpice allows you to change circuit or source parameters easily, you can explore the effects that these changes have on circuit variables more easily than with a calculator. This kind of exploration is much like the experience that you can obtain doing circuit laboratory experiments and is invaluable in helping you to understand how circuits work.

The digits before the decimal point of each problem number in this chapter identify the corresponding chapter in Cunningham and Stuller. The digits following the decimal point give this manual's sequence number for each problem. This manual includes a DOS-format floppy disk that has all of the example and problem netlists. The file name of each netlist relates to the problem number in this manual. For example, file PR4_1.CIR is the netlist for Problem 4.1, the first problem in this manual that relates to Chapter 4 in Cunningham and Stuller. Each problem file is nearly complete. You need to fill in missing information for each netlist. Netlist solutions also are on the disk. The solution files have the same name as the problem file, but the letter "s" appears after the name and before the CIR extension. For example, the solution file for problem file PR4_1.CIR is PR4_1s.CIR.

If you use a Macintosh to run PSpice, convert the contents of the floppy disk to Macintosh format. Use the Apple File Exchange program, which exists on your Macintosh System Disks, to do this conversion. For details about this process, see Secs. 1.1.2 and 1.3.3.

## Problems for Textbook Chapter 4

The following problems relate to Chapter 4 of Cunningham and Stuller and illustrate PSpice solutions of circuit problems that complement solutions using node voltage or mesh current analysis. Some of these problems show how network reduction simplifies the simulation or analysis. Others explain how superposition applies for circuits having more than one independent source. Some examples demonstrate how to use nested DC sweeps or alphanumeric node names.

## 4.1 Nodal Analysis with Independent Current Sources

(Chapter 4, Problem 3)
For the circuit shown in Fig. P4.1, obtain a PSpice solution for all of the node voltages and voltage $v_x$. Verify from the solution that

$$v_x = v_2 - v_3$$

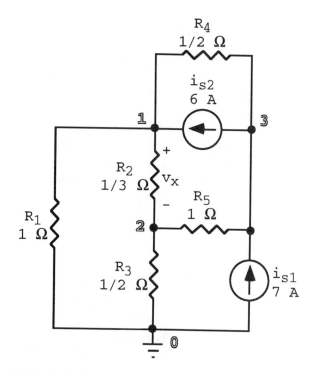

FIGURE P4.1

## 4.2 Nodal Analysis with Dependent Source

(Chapter 4, Problem 4)

Write a PSpice netlist for the circuit of Fig. P4.2 to obtain values for all node voltages and currents $i_a$, $i_b$, $i_c$, and $i_x$. Notice that the circuit in Fig. P4.2 has two null-valued voltage sources. One measures the control current $i_a$, and the other measures current $i_x$. Examine the output file to see that the solution values agree with Kirchhoff's current law.

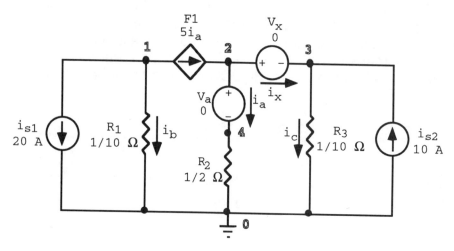

FIGURE P4.2

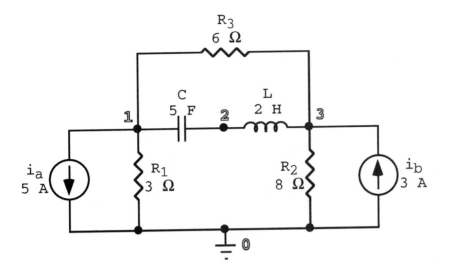

FIGURE P4.3

## 4.3 Nodal Analysis with Inductance and Capacitance

(Chapter 4, Problem 5)
Write a PSpice netlist for the circuit shown in Fig. P4.3. Measure all device currents, all node voltages, and the voltage $v_c$ across capacitance C. Why is the capacitance voltage nonzero when the inductance current is zero?

## 4.4 Nodal Analysis with Voltage Sources

(Chapter 4, Problem 2)
Write a PSpice netlist for the circuit shown in Fig. P4.4. Measure the node voltages and all of the currents. Verify that the currents agree with Ohm's law, the node voltages, and Kirchhoff's current law.

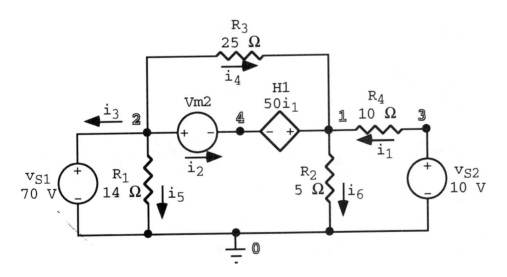

FIGURE P4.4

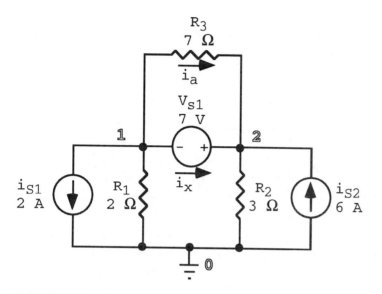

FIGURE P4.5

## 4.5 Nodal Analysis with Supernode

(Chapter 4, Problem 6)

a) Write a PSpice netlist for the circuit shown in Fig. P4.5. Have the simulation measure the values of currents $-i_x$ and $i_a$.

b) Change the value of resistance $R_3$ to 10 $\Omega$. Does this change affect the value of $-i_x$ or $i_a$?

c) Place an asterisk (*) in front of the statement for resistance $R_3$ to remove $R_3$ from the circuit, and remove reference to current $i_a$. What effect does this change have on the current $-i_x$?

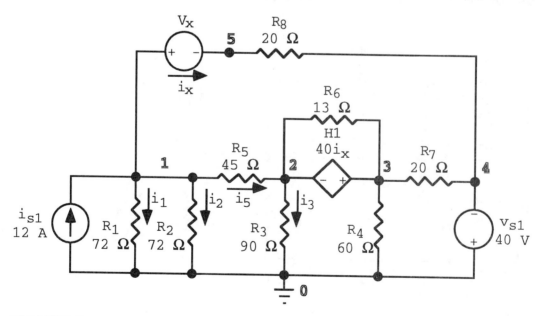

FIGURE P4.6

## 4.6 Nodal Analysis and Network Reduction

(Chapter 4, Problem 9)

a) Do a PSpice node analysis of the circuit shown in Fig. P4.6 to determine the values of all node voltages and currents $i_1$, $i_2$, $i_3$, and $i_5$.

b) Reduce resistances $R_1$ and $R_2$ to a single equivalent resistance. Run PSpice again, replacing $R_1$ and $R_2$ with this single equivalent resistance. Obtain all node voltages and currents $i_3$, $i_5$, and the current through the equivalent resistance. Does the current through the equivalent resistance equal the sum of the currents $i_1$ and $i_2$ from part a?

c) Vary resistance $R_6$. Does the value of resistance $R_6$ affect the node voltages or currents that you observe? Measure the current in resistance $R_6$ and in the CCVS H1. Explain why these currents change values as resistance $R_6$ changes.

## 4.7 Transfer Ratios

(Chapter 4, Problem 11)

For the circuit shown in Fig. P4.7 determine the voltage ratios $v_e/v_s$ and $v_c/v_s$, the current gains $i_e/i_b$ and $i_c/i_b$, the power gains $v_e i_e/v_s i_b$ and $-v_c i_c/v_s i_b$, and the input resistance $v_s/i_b$. To make these measurements, use the following techniques:

a) Use a 1-V sized voltage source $v_s$.

b) Use four VALUE-type VCVS devices to measure the current and power gains.

c) Use the TF command to measure the input resistance.

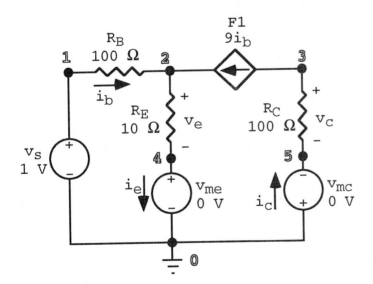

FIGURE P4.7

---

Observe from the output file that each power gain equals the product of an appropriate voltage and current gain.

## 4.8 Mesh Currents

(Chapter 4, Problem 25)

For the circuit shown in Fig. P4.8, write a netlist to solve for the mesh currents $i_1$, $i_2$, and $i_3$. Also measure the voltages at nodes 2 and 5.

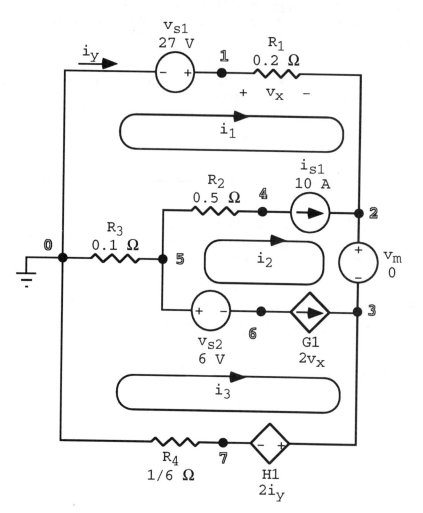

FIGURE P4.8

## 4.9 Mesh Analysis

(Chapter 4, Problem 26)

Write a PSpice netlist for the circuit shown in Fig. P4.9 to evaluate the mesh currents $i_1$, $i_2$, $i_3$, and $i_4$. Measure the current in the VCCS G1and compare this value with the difference between $i_2$ and $i_3$.

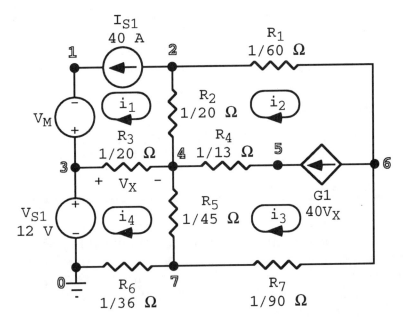

FIGURE P4.9

## 4.10 Nodal Analysis and Network Reduction

(Chapter 4, Problem 45)

a) Write a PSpice netlist to solve the circuit shown in Fig. P4.10 for all node voltages, the node voltage difference $v_x$, and currents $i_x$ and $i_y$.

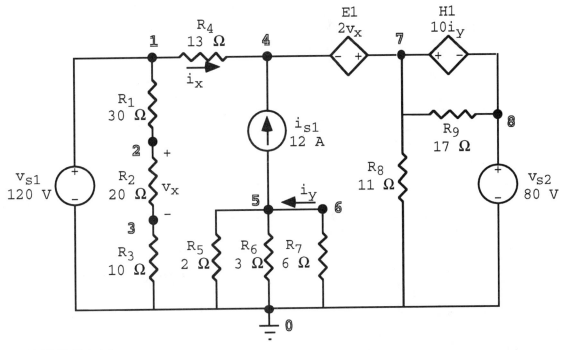

FIGURE P4.10

b) Using current and voltage divider rules, calculate $v_x$ and $i_y$. With these values, now calculate the voltage value of the E1 and H1 source. Identify the node-voltage difference across resistance $R_4$ and calculate current $i_x$. Compare this solution with the PSpice simulation.

## Problems for Textbook Chapter 5

The problems in this section relate to Chapter 5 of Cunningham and Stuller and illustrate linearity, superposition, Thévenin's theorem, and Norton's theorem.

## 5.1 Node Analysis and Superposition

(Chapter 5, Problem 1a)

Write a PSpice netlist to solve for the current $i_Y$ in the circuit shown in Fig. P5.1. Obtain the solution using superposition, finding the components of $i_Y$ due to $i_{S1}$, $i_{S2}$, and $v_{S1}$ alone. Superimpose these three values to obtain the value of $i_Y$ when all three sources are active. Compare this superposition value with a value using PSpice with all three sources on.

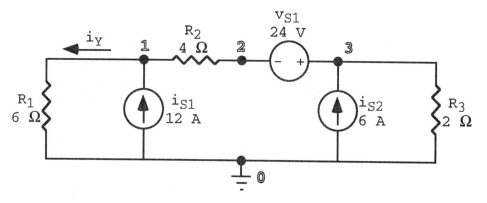

FIGURE P5.1

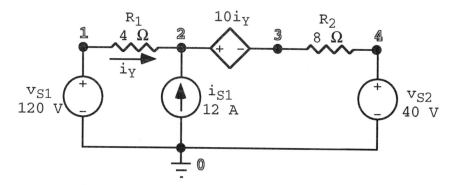

FIGURE P5.2

## 5.2 Node Analysis with Controlled Source and Superposition

(Chapter 5, Problem 1b)
Write a PSpice netlist to solve for the current $i_Y$ in the circuit shown in Fig. P5.2. Obtain the solution using superposition, finding the components of $i_Y$ due to $i_{S1}$, $i_{S2}$, and $v_{S1}$ alone. Superimpose these three values to obtain the value of $i_Y$ when all three sources are active. Compare this superposition value with a value using PSpice with all three sources on.

## 5.3 Design with Superposition

(Chapter 5, Problem 3)
For the circuit shown in Fig. P5.3

   a) Write a netlist and simulate the circuit with four values of $v_a$ and $v_b$ using the nested version of the DC command. Make each voltage source step from 0 V to 1 V with a 1-V step. Determine from the output file the values of the ratios of $v_o/v_a$ with $v_b$ equal to zero and $v_o/v_b$ with $v_a$ equal to zero. From these results conclude that

$$v_o = K(v_a + v_b)$$

   What is the value of the constant K?

   b) In a separate netlist, set $v_a$ to 1 V and $v_b$ to 0 V. Using a DC command, sweep resistance $R_a$ from 10 $\Omega$ to 100 $\Omega$ in 1-$\Omega$ steps. Include a PROBE statement in your netlist and use the DAT file to plot $v_o$ vs. $R_a$. Move the cursor to the point on this plot where the $v_o$ value is twice the value when $R_a$ is 40 $\Omega$. What is the $R_a$ value to achieve this gain?

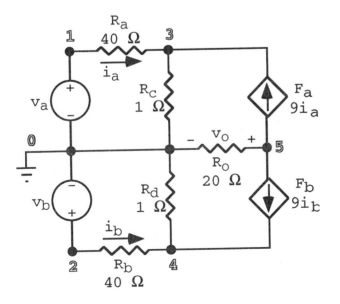

FIGURE P5.3

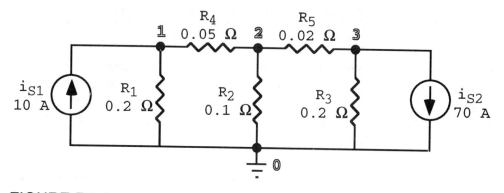

FIGURE P5.4

## 5.4 Node Analysis and Network Transformation

(Chapter 5, Problem 6)
For the circuit shown in Fig. P5.4

   a) Write a netlist and use PSpice to solve the circuit. Determine the three node voltages and the currents in all five resistances.

   b) Write a second netlist and solve for the same voltages and all five resistance currents after transforming the current source $i_{S1}$ and resistance $R_1$ and the current source $i_{S2}$ and resistance $R_3$ into equivalent voltage source and resistance circuits. Why are the currents in resistances $R_1$ and $R_3$ different from their values in the first circuit, although the other resistance currents remain the same?

## 5.5 Thévenin Equivalent of a Circuit

(Chapter 5, Problem 9c)
For the circuit shown in Fig. P5.5

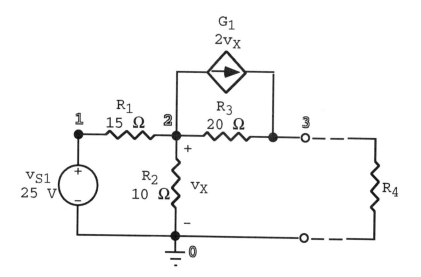

FIGURE P5.5

a) Write a netlist description to find the Thévenin equivalent open-circuit voltage at node 3 with respect to the reference node with $R_4$ not in the circuit. Include a TF statement to measure the Thévenin circuit parameters directly.

b) Write a second netlist to find the short-circuit current by attaching a null-valued voltage source from node 3 to node 0. Compare the Thévenin equivalent using the open-circuit voltage and short-circuit current with the values from the TF command.

c) Write a third netlist to compare the Thévenin equivalent circuit with the original circuit as the resistance $R_4$ in the first circuit varies from 25 $\Omega$ to 500 $\Omega$ in 5-$\Omega$ steps. Use a resistance model to vary load resistance $R_4$ in the original circuit and in the Thévenin equivalent circuit. The Thévenin equivalent circuit shares the reference node with the first circuit, but is otherwise a separate circuit that you include in the same netlist. Include a PROBE statement, and limit the output in the DAT file to only the variables that you need. Plot the output power of both circuits [V(3)*I(R4) for the original circuit] using the Probe application.

In your solution set the precision to six digits.

## 5.6 Maximum Power to a Load Circuit

(Chapter 5, Problem 17)

For the circuit shown in Fig. P5.6, write a netlist to obtain a Probe plot of the power in the load circuit as a function of parameter r. Vary parameter r from 1 V/A to 10 V/A in 0.2 V/A steps. Set the cursor at the maximum power to determine the value of r that gives maximum power and this maximum power.

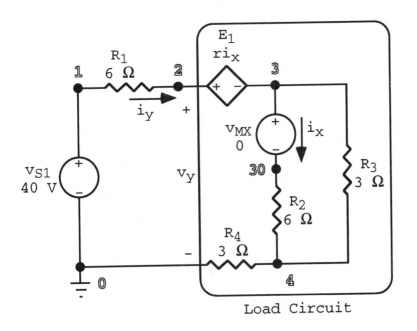

Load Circuit

FIGURE P5.6

Restrict output in the DAT file to the load circuit voltage $v_y$ and load current $i_y$. Notice the use of the VALUE form of E source here, because parameter r cannot appear as a value in an H-source statement. The analog behavior form of an E statement can use a parameter in the expression that defines the voltage source value.

## 5.7 Norton Equivalent and Maximum Load Power

(Chapter 5, Problem 28)

Simulate the circuit shown in Fig. P5.7 to obtain data so that with Probe you can plot the currents in resistance $R_1$ and $R_2$ and the power in resistance $R_2$ as resistance $R_2$ varies from 1 kΩ to 12 kΩ in 200 Ω steps. Vary resistance $R_2$ using either a parameter or a resistance model.

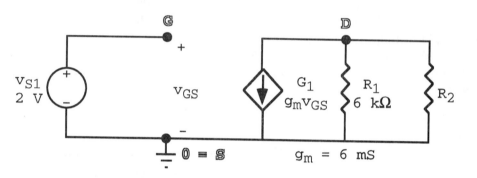

FIGURE P5.7

## Problems for Textbook Chapter 6

The problems in this section relate to Chapter 6 of Cunningham and Stuller and show how to model op amps and op-amp application circuits using PSpice.

## 6.1 Op-Amp Noninverting Amplifier Application Circuit

(Chapter 6, Example 1 )

The circuit in Fig. P6.1 is a noninverting op-amp application circuit. The model of the op amp in the circuit in Fig. P6.1 includes only the large but finite voltage gain $A_d$. Using the VALUE form of E source so that you can vary the gain $A_d$, simulate the circuit with 5 points per decade as $A_d$ varies from 100 to $10^6$. Since the source voltage $v_{S1}$ is 1 V, the output directly measures the application circuit's voltage gain A. Using Probe, plot the voltage gain A as a function of the op-amp gain $A_d$. Also, plot the op-amp input voltage as a function of the op-amp gain $A_d$. The gain error due to finite op-amp gain $A_d$ is

$$\text{Gain Error Percent} = \frac{A - \left(1 + R_2 / R_1\right)}{1 + R_2 / R_1} \times 100 = 10A - 100$$

Plot the gain error in percent as a function of the op-amp gain $A_d$. Maintain six digits of precision.

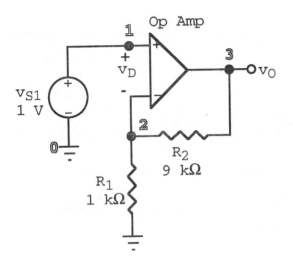

FIGURE P6.1

## 6.2 Op-Amp Inverting Amplifier Application Circuit

(Chapter 6, Example 2 )

Repeat Problem 6.1 for the inverting amplifier application circuit shown in Fig. P6.2. For this circuit, the gain error due to finite op-amp gain $A_d$ is

$$\text{Gain Error Percent} = \frac{A - R_2/R_1}{-R_2/R_1} \times 100 = -10A + 100$$

Plot the gain error in percent as a function of the op-amp gain $A_d$. Maintain six digits of precision.

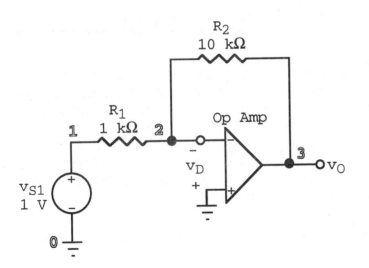

FIGURE P6.2

## 6.3 Effect of Op-Amp Input Resistance on Noninverting Circuit

(Chapter 6, Example 1 )

This problem explores the effect of op-amp input resistance on the noninverting op-amp application circuit's voltage gain. Insert an input resistance $R_{in}$ between the input terminals of the op amp in Fig. P6.1 (between nodes 1 and 2). Sweep this resistance from 10 k$\Omega$ to 100 M$\Omega$ with 5 points per decade. Obtain a Probe plot of the voltage gain and the gain error. Keep the op-amp difference-mode gain equal to $10^5$ for this PSpice experiment.

## 6.4 Thévenin Equivalent of Inverting Op-Amp Circuit

(Chapter 6, Example 2)

For the circuit shown in Fig. P6.4 using PSpice:

   a) Measure the Thévenin open-circuit voltage.

   b) Measure the Norton short-circuit current.

   c) Set source $v_{S1}$ to zero and apply a 1-A current source from the reference node to output node 4 to measure the Thévenin equivalent resistance.

   d) Obtain the Thévenin equivalent circuit parameters using the TF command.

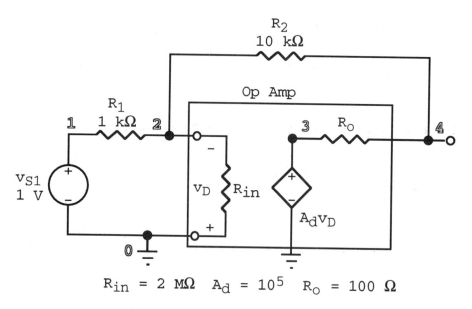

$$R_{in} = 2 \text{ M}\Omega \quad A_d = 10^5 \quad R_o = 100 \ \Omega$$

FIGURE P6.4

## 6.5 Difference Amplifier Circuit

(Chapter 6, Exercise 3)

The circuit shown in Fig. P6.5 is a difference amplifier circuit that gives an output proportional to the difference between the two input voltages $v_1$ and $v_2$ when the resistance ratio $R_2/R_1$ equals the resistance ratio $R_b/R_a$. In the

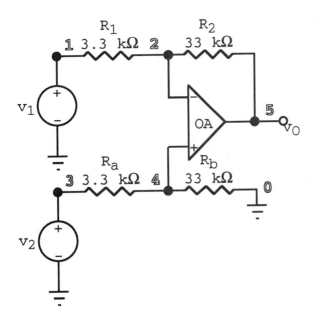

FIGURE P6.5

following, model the op amp using an E-type VCVS having a gain equal to $2 \times 10^5$.

a) Using a nested DC sweep, find the ratio of $v_O$ to $v_1$ with $v_2$ equal to zero and the ratio of $v_O$ to $v_2$ with $v_1$ equal to zero. Include the source currents in the output file and calculate the resistance seen by each source.

b) With $v_2$ equal to 2 V, vary $v_{S1}$ from -5 V to 5 V using 0.5-V steps and use Probe to plot the output voltage as a function of $v_1$. Does the output equal zero for the value of $v_1$ that you expect? If the op amp operates with ±15 V DC supplies and the output limits at ±14 V, how will the output voltage as a function of the input voltage change?

## 6.6 Tee-Feedback-Circuit Gain

(Chapter 6, Exercise 5)

The circuit shown in Fig. P6.6 gives large voltage gain using medium-sized resistance values. Source E1 in the circuit models the op amp's large but finite voltage gain $A_d = 2 \times 10^5$ V/V. Vary resistance $R_4$ linearly from 200 Ω to 2 kΩ in 100-Ω steps. Plot the voltage gain from $v_{S1}$ to both node 3 and to the output node 4 as a function of $R_4$. Use the cursor to find the value of $R_4$ that gives a voltage gain at the output of -100 V/V.

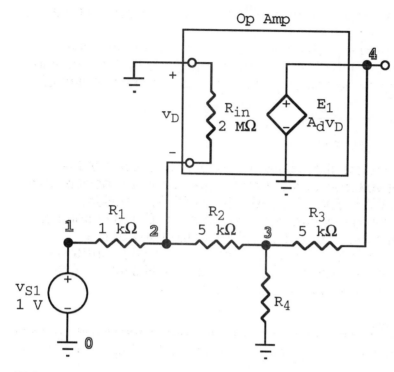

FIGURE P6.6

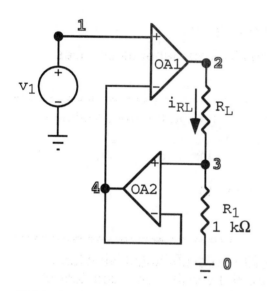

FIGURE P6.7

## 6.7 Voltage-to-Current Converter

The op-amp application circuit shown in Fig. P6.7 is a voltage-to-current converter circuit. The op amps force the voltage at node 3 to equal the source voltage $v_1$. Therefore, the current $i_{RL}$ is $v_1/R_1$, independent of the value of the load resistance $R_L$.

   Write a PSpice netlist for this circuit that varies the load resistance from 1 k$\Omega$ to 5 k$\Omega$ in 1-k$\Omega$ steps with the DC command. Change the source voltage

$v_1$ from 1 V to 5 V in 1-V steps using the STEP command. Model the op amps using an E source with a gain of 200 kV/V.

Include a PROBE statement. Using Probe, plot the current $i_{RL}$ as a function of the load resistance $R_L$ with $v_1$ as parameter. Verify that the voltages at node 3 and at node 4 equal the source voltage $v_1$ and are independent of the load resistance $R_L$.

## Problems for Textbook Chapter 7

PSpice can model the waveforms that you learn about in Chapter 7 of the textbook. The independent voltage source V and current source I devices can have transient forms using PWL, EXP, PULSE, SIN, or SFFM transient specifications. In addition, the behavioral modeling forms of E and G sources provide another way to describe transient waveforms using any of the mathematical functions in Table 2.8. Voltage (S) and current (W) controlled switches provide another way to create transients in electrical circuits. The following problems illustrate these features of PSpice.

### 7.1 Unit-Step Functions

With Probe, use a PWL transient specification to plot the following voltage-step functions for $0 \text{ s} \le t \le 10 \text{ s}$:

   a) $v_A(t) = 2u(t-5) \text{ V}$

   b) $v_B(t) = u(t) - u(t-2) + u(t-4) - u(t-6) + u(t-8)$

Use a time interval of 1 ms to establish apparently instantaneous changes.

### 7.2 Unit-Ramp Functions

With Probe, use the PWL transient specification to plot the following voltage-ramp functions for $0 \text{ s} \le t \le 10 \text{ s}$:

   a) $v_A(t) = 2r(t-5) \text{ V}$

   b) $v_B(t) = r(t) - r(t-1) - r(t-4) + r(t-6) + r(t-9)$

### 7.3 Exponential Functions

   a) For time between 0 and 10 s, simulate an exponential waveform having a 2-V initial value, rising toward 8 V with a time constant of 2 s after a delay of 1 s. Use the Probe functions d(x) and s(x) to display the derivative and integral of the simulation waveform.

   b) Repeat Part a for a waveform having an initial 8-V value, falling toward 2 V, and keeping the delay and time constant the same.

   c) Simulate an exponentially-rising waveform beginning at 1 s, rising from 1 V, and heading toward 9 V with a 0.5-s time constant. At 5 s, the waveform begins to return to 1 V with a recovery time constant of 1 s. This waveform has the standard form given in Sec. 2.4.2. Examine the waveform using Probe and be sure that you understand the meaning of each parameter.

## 7.4 Pulse Functions

For time between 0 and 100 μs, simulate a periodically repetitive pulse waveform having a delay time of 1 μs, rise and fall times of 1 μs, a pulse width of 2 μs, and a period of 10 μs. The starting voltage is 1 V and the peak value is 5 V. Use Probe to display the waveform. Superimpose the Probe time average function avg(x) of this waveform on your display. Move the cursor to the value of the time average display at 100 μs. Is the value here consistent with the average of the pulse waveform? Repeat for a waveform having a starting value of 5 V and a peak value of 1 V.

## 7.5 Sinusoidal Function

For time between 0 and 10 ms, simulate a 1-ms delay, damped sinusoid that has a 1-V offset, a 6-V amplitude, a 1-kHz frequency, a damping factor equal to 0.5 krad/s, and a phase shift of 45°. Use Probe to display the waveform. Superimpose on this waveform the positive and negative exponential envelope function

$$\text{Envelope Function} = 1 \pm 6e^{-0.5\,\text{k}*(\text{time}-1\text{m})}\ \text{V}$$

Use the Probe cursor to measure the successive positive and negative maximum values. Compare these values to ones that you compute using the waveform's function.

## 7.6 Full-Wave-Rectified Sine-Wave Function

Use the behavioral modeling E source and the abs(x) function to simulate ten periods of a full-wave rectified sinusoid having a peak amplitude of 10 V and a frequency of 1 kHz. Use Probe to display the waveform. Superimpose the average value on the waveform using the avg(x) and rms(x) Probe functions. Do these values approach the value that you expect?

## 7.7 Thévenin Equivalent of a Switched Circuit

For the circuit shown in Fig. P7.7, switch S1 closes at t = 5 s. Write a PSpice netlist and use Probe to display the open-circuit voltage and the Thévenin equivalent resistance seen by current source iₛ between nodes 4 and 0 as a function of time for 0 ≤ t ≤ 10 s. Make the voltage source vₛ step from 0 to 18 V at 8 s, and the current source iₛ step from 1 to 0 A at 8 s. Connect a voltage source v_C between node 5 (not shown) and node 0 to control the switch. Have this control source step from 0 to 1 V at 5 s to make the switch turn on at 5 s. Use RON = 1 mΩ and ROFF = 1 MΩ for the switch to behave as a short circuit when on and an open circuit when off. Observe the voltage at node 4 using Probe. From the changing values that you see, what is the value of the Thévenin resistance before and after the switch closes? What is the open-circuit voltage after the switch closes?

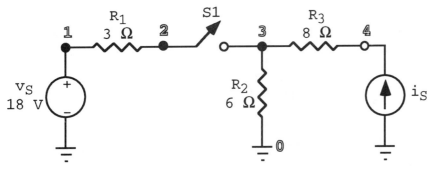

FIGURE P7.7

## Problems for Textbook Chapter 8

Textbook Chapter 8 describes transient analysis of first-order circuits. These circuits can have nonzero initial energy conditions due to currents in inductances or voltages across capacitances due to voltage or current sources that exist prior to a switching event. These sources can connect or disconnect from the circuit or change their values abruptly at the initial moment. The initial moment usually is taken to be the origin of time. Because SPICE does not deal with time values less than zero and the optional transient specification UIC (Use Initial Condition) applies any initial condition values for capacitances or inductances to time zero, transient simulations usually start at time zero. If a problem specifies initial conditions at any time other than zero, you can translate time to make the initial time be zero. Frequently, transient problems specify the initial conditions to be at time zero. Although many transient problems specify switching actions that occur at time zero, a simulation of this sort does not require use of either current- or voltage-controlled switches because the circuit that comes into being at time zero persists thereafter.

The simulations in this section show how to run transient simulations using SPICE. The command that implements a transient simulation is the TRAN command, often with the UIC option. Initial condition specifications for capacitances or inductances define the initial energy state of these elements. Measurement of voltages and currents uses the PRINT TRAN command with specification of variables. The PLOT TRAN command gives a line plot of circuit variables in the OUT file. Since Probe plots circuit variables better than the line plots of the PLOT command and also has other useful features, we will not use the PLOT command.

## 8.1 RC Source-Free Circuit

(Chapter 8, Problem 4)

Simulate the circuit shown in Fig. P8.1 with $\beta$ equal to 4, 9, and 19 using the STEP command. Since PSpice parameters cannot appear in the simple form for controlled sources, use the VALUE form of the G-type source to simulate the CCCS. Use Probe to display each of the following items as a function of time for the three values of $\beta$:

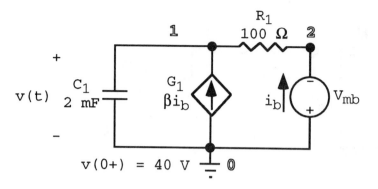

FIGURE P8.1

a) The capacitance voltage.

b) The derivative of the capacitance voltage using the d(x) PSpice function. From this plot compute the time constant using the fact that the magnitude of the initial slope equals the initial voltage divided by the time constant.

c) The integral of the capacitance voltage using the s(x) PSpice function. From this plot compute the time constant using the fact that the integral of the voltage for time between zero and infinity equals the product of the initial voltage and the time constant.

d) The negative of the capacitance value times the capacitance voltage divided by the capacitance current. Why does this computation display the time constant?

e) The integral of the negative of the capacitance voltage times the capacitance current. Why are the final values of each curve the same, regardless of the value of β?

Use <t_Print> and <t_Final> values in the TRAN command so that the shortest time constant solution has five display points in one time constant and the longest time constant solution display extends over five time constants.

## 8.2 RL  Source-Free  Circuit

(Chapter 8, Problem 8)

For the circuit shown in Fig. P8.2

a) Replace the inductance $L_1$ with a 1-A current source and simulate the resulting circuit to determine the Thévenin resistance seen by inductance $L_1$.

b) Remove the current source and reinstall the inductance $L_1$. Now, simulate the circuit and observe the transient solution for the inductance current i(t). Use <t_Print> and <t_Final> values in the TRAN command so that the solution has approximately five display points in one time constant and the display extends over approximately five time constants.

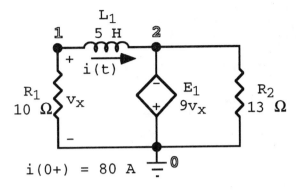

$i(0+) = 80$ A

FIGURE P8.2

## 8.3 RC Circuit Switching Transient

(Chapter 8, Example 9)

After being closed for a long time, the switch S1 in Fig. P8.3 opens at t = 0 and closes again at t = 0.1 s.

a) Simulate this circuit with PSpice and use Probe to plot current i(t) as a function of time. Superimpose the solutions

$$i(t) = 12 + 8e^{-10t} \text{ A} \quad 0 \le t \le 0.1\,\text{s}$$

and

$$i(t) = 20 - 0.5057e^{-6(t-0.1)} \text{ A} \quad t \ge 0.1\,\text{s}$$

to compare the PSpice simulation with the solution.

b) Revise the netlist of Part a so that both resistances $R_1$ and $R_2$ can change using a resistance model. Use the STEP command with the LIST option so that these resistances have values equal to 0.9, 1.0, and 1.1 times their nominal values in Fig. P8.3. Plot current i(t) as a function of time for these three cases using Probe.

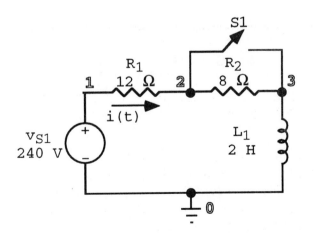

FIGURE P8.3

c) Revise the netlist of Part a so that the time when the switch closes again can change. Use the STEP command with the LIST option so that these times have values equal to 50 ms, 100 ms, and 150 ms. Plot current i(t) as a function of time for these three cases using Probe.

## 8.4 Complete Response of RC Circuit with Two Sources

(Chapter 8, Exercise 17)
For the circuit shown in Fig. P8.4

$$i_{s_1}(t) = 4e^{-t} \text{ A} \qquad t \geq 0$$

$$i_{s_2}(t) = 2 \text{ A} \qquad t \geq 0$$

Write a netlist and simulate the circuit with PSpice. Use Probe to plot the voltage v(t) as a function of time. Superimpose on this plot the solution

$$v(t) = 4 + 12e^{-t} - 10e^{-3t} \text{ V}$$

from Cunningham and Stuller. Also plot the particular solution. What is the time duration before the particular solution and the complete solution merge? (Note: You can replace the two sources with one equal to the sum of the two in the circuit diagram. However, to simulate the circuit in SPICE you have to use two sources, because SPICE ignores a DC value when the source also has a transient specification.)

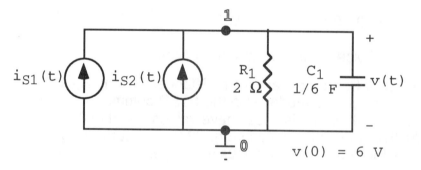

FIGURE P8.4

## 8.5 RC Circuit Response to a Sinusoidal Source

(Chapter 8, Example 11)
With

$$v_s(t) = \frac{15}{\sqrt{2}}\cos(3t) \text{ V}$$

the circuit shown in Fig. P8.5 has a steady-state response

$$v_p(t) = 9\cos(3t + 8.13°) \text{ V}$$

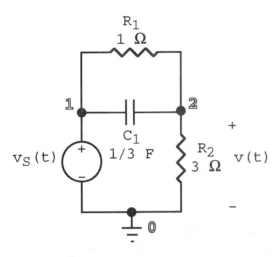

FIGURE P8.5

Use PSpice and Probe to plot the voltage v(t) as a function of time. Superimpose the particular solution on this plot. You will see that the complementary solution disappears in less than half of the particular solution's period. Remember to express the 8.13° phase shift in radians for the Probe expression.

## Problems for Textbook Chapter 9

Textbook Chapter 9 shows how to solve transient probems for second-order circuits. These circuits have two independent energy storage elements.

## 9.1 Parallel RLC Source-Free Circuit

(Chapter 9, Problem 1)

For the circuit shown in Fig. P9.1

> a) Write a netlist and use Probe to obtain plots of inductance current i and capacitance voltage v as a function of time. Vary the resistance using one point per octave from 0.25 to 4 times the resistance value $R_{Critical}$ that gives critical damping. Select <t_Print> to display at least 20 points per period for any underdamped case and at least five points occur for the shortest time constant of any overdamped case. Choose <t_Final> on the order of 5 times the longest time constant or reciprocal damping coefficient. To maintain computation accuracy select <t_Ceil> equal to 1/200 of the <t_Final> value.
>
> b) Plot the inductance energy and capacitance voltage as a function of time using Probe for all five values of resistance.
>
> c) Selecting the data for $R = 0.25R_{Critical}$, plot the inductance energy and the capacitance energy as functions of time. Explain why the capacitance energy equals zero when the inductance energy is a maximum.

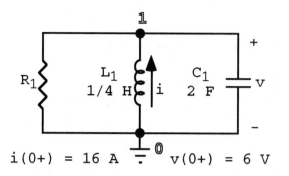

$i(0+) = 16 \text{ A}$    $v(0+) = 6 \text{ V}$

FIGURE P9.1

Repeat this problem with $i(0+) = 0$ and $v(0+) = 6$ V and with $i(0+) = 16$ A and $v(0+) = 0$ to see how the solutions change as the initial conditions change.

## 9.2 Series RLC Source-Free Circuit

(Chapter 9, Problem 7)

Repeat Problem 9.1 for the series RLC circuit shown in Fig. P9.2, except in Part c explain why the inductance energy equals zero when the capacitance energy is a maximum.

Repeat this problem with $i(0+) = 0$ and $v(0+) = 16$ V and with $i(0+) = 6$ A and $v(0+) = 0$ to see how the solutions change as the initial conditions change.

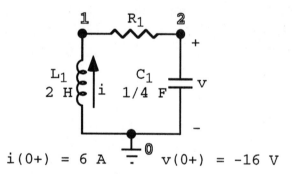

$i(0+) = 6 \text{ A}$    $v(0+) = -16 \text{ V}$

FIGURE P9.2

## 9.3 RLC Circuit Response to an Exponential Source

(Chapter 9, Example 7)

For the circuit shown in Fig. P9.3, write a netlist and use Probe to plot voltage $v_1$ as a function of time. Assume that the initial capacitance voltage and inductance current are zero. Superimpose the particular solution

$$v_{1p}(t) = -30e^{-3t}u(t) \text{ V}$$

onto this plot. Do the complete solution and the particular solutions merge after five time constants?

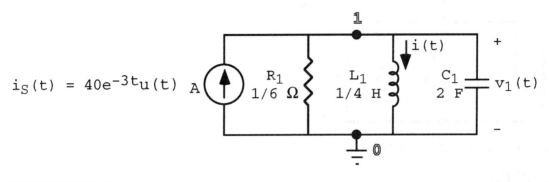

$$i_S(t) = 40e^{-3t}u(t) \text{ A}$$

FIGURE P9.3

## 9.4 RLC Circuit Switching Transient

(Chapter 9, Problem 20)

The switch S1 in Fig. P9.4 opens after being closed for a long time and then closes again at t = 1 s. Simulate this circuit and use Probe to plot capacitance voltage v(t) and inductance current i(t) as functions of time for $0 \le t \le 2$ s. Calculate the capacitance voltage v(t) and the inductance current i(t) waveforms. Superimpose these calculation values on the Probe plots. Using the inductance-current waveform for the first 1-s interval, measure the fast-transient and slow-transient damping factors and the amplitude of these transient components. The interchange of resistance values of $R_1$ and $R_2$ from the values in Problem 9.20 of Cunningham and Stuller gives an underdamped solution when the switch closes.

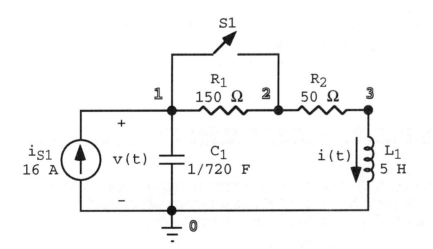

FIGURE P9.4

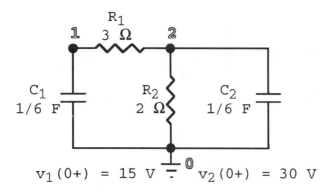

$$v_1(0+) \ = \ 15 \ V \qquad v_2(0+) \ = \ 30 \ V$$

FIGURE P9.5

## 9.5 Two-Capacitance Source-Free Circuit

Simulate the circuit shown in Fig. P9.5. Use Probe to plot voltages $v_1$ and $v_2$ as functions of time. Also plot the capacitance energies as functions of time. Make <t_Print> as small as 1/5 the smallest time constant and <t_Final> 5 times the longest time constant.

## Problems for Textbook Chapter 10

The problems in this section show how to simulate AC sinusoidal steady-state circuits. To determine phasor solution variables with SPICE you use the AC command. With this command SPICE finds phasor solutions for the circuit that results from a linearization of the original circuit about the operating point of the circuit. For a linear circuit there is no difference between this linearization and the circuit with all DC components of sources set to zero.

## 10.1 RLC Series Circuit at a Single Frequency

(Chapter 10, Example 4)

Use the AC command to do a phasor circuit analysis for the time-domain circuit in Fig. P10.1. Print the magnitude and phase of the source current $i_s(t)$ and the voltage $v(t)$. Maintain six digits of accuracy in the output values.

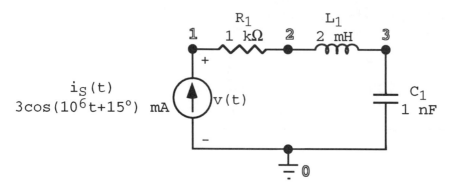

FIGURE P10.1

## 10.2 RLC Series-Circuit Frequency Response

(Chapter 10, Example 4)

Repeat the simulation of the circuit in Fig. P10.1 letting the frequency sweep at 20 points per decades from 10 kHz to 1 MHz. Include the PROBE command in the netlist. Using Probe, plot the magnitude of the resistance, inductance, and capacitance voltages and the source voltage as a function of frequency. Also plot the phase of these variables as functions of the frequency. Do these plots display what you expect to see? Also, plot the magnitude and phase of the ratio of V(1)/I(Is1) using the Probe functions M(x) and P(x). From your observations you can conclude that Probe does complex arithmetic when evaluating expressions. Use Probe's cursor function to locate the frequency where the magnitude of V(1) is a minimum and the phase angle equals zero. How does this frequency value relate to the values of the circuit parameters?

## 10.3 RLC Series Circuit Time-Domain Solution

(Chapter 10, Example 4)

For the circuit of Fig. P10.1 do a simulation using the transient SIN function to represent current source $i_{S1}(t)$ with frequency values of 10 kHz, 112.540 kHz, and 1 MHz with the circuit in periodic steady-state operation. (Set the initial inductance current and capacitance voltage to their correct values.) Using Probe, plot the resistance, inductance, and capacitance voltage as functions of time at each frequency. Use Probe's cursor function to measure the magnitude and phase of each waveform.

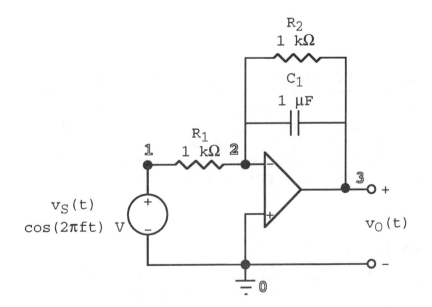

FIGURE P10.4

## 10.4 Inverting Op-Amp RC Circuit Frequency Response

Simulate the circuit shown in Fig. P10.4 using the AC command for frequencies ranging from 10 to 100 kHz with 20 points per decade. Plot the magnitude and phase of the output voltage $v_O(t)$ with Probe. Use Probe's cursor function to measure the frequency where the magnitude equals $1/\sqrt{2}$ and the phase shift becomes -225°. Model the op amp using a VCVS source having a gain of $10^5$ V/V.

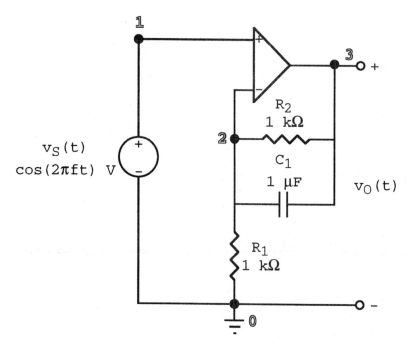

FIGURE P10.5

## 10.5 Noninverting Op-Amp RC Circuit Frequency Response

Simulate the circuit shown in Fig. P10.5 using the AC command for frequencies ranging from 10 Hz to 100 kHz with 20 points per decade. Plot the magnitude and phase of the output voltage $v_O(t)$ with Probe. Compare the low- and high-frequency magnitude and phase values to those that you expect. Use Probe's cursor function to measure the frequency where the phase becomes a minimum.

### Problems for Textbook Chapter 11

The problems in this section give PSpice solutions to compare with analytical solutions that use series, parallel, and divider rules and node and mesh analysis in the phasor domain.

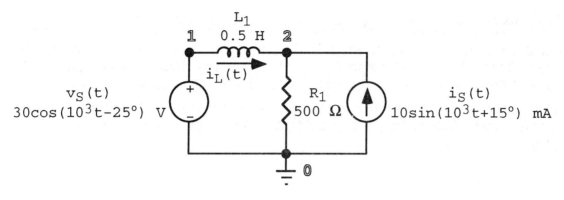

FIGURE P11.1

## 11.1 RL Series Circuit at a Single Frequency

(Chapter 11, Example 4)

For the circuit shown in Fig. P11.1, use the AC command to determine the magnitude and phase of the complex current $I_l$ that represents the time-domain inductance current $i_L(t)$. Compare the PSpice solution with the solution in Example 11.4 of Cunningham and Stuller. Maintain six digits of precision.

## 11.2 Phasor Thévenin Equivalent of a Circuit

(Chapter 11, Example 19)

For the time-domain circuit shown in Fig. P11.2, use PSpice to determine the phasor Thévenin equivalent open-circuit voltage in polar form and Thévenin equivalent impedance in rectangular form at terminals a-b.

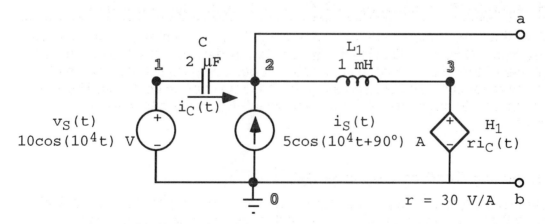

FIGURE P11.2

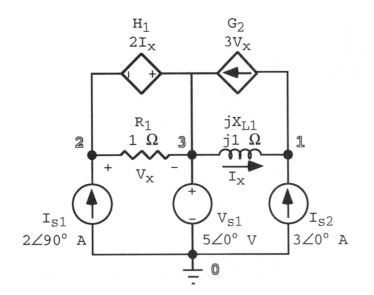

FIGURE P11.3

## 11.3 Circuit with   Given  Impedances

(Chapter 11, Problem 26)

Use PSpice to determine phasor current $I_x$, voltage $V_x$ and the node voltages at nodes 1 and 2 for the circuit shown in Fig. P11.3. Find the values in polar form to 6 digits of precision.

## Problems for  Textbook  Chapter  12

The problems in this section show how to calculate or plot  instantaneous power, real and reactive power, and power factor using PSpice and Probe. The netlists here use the AC, TRAN, STEP, and MODEL commands, the AC source specification, and the SIN function transient specification.

## 12.1  Power  in  a  Series  RC  Circuit

(Chapter 12, Exercise 2)

For the circuit shown in Fig. P12.1, the resistance $R_1$ is

$$R_1 = 10\cos(45°) = \frac{10}{\sqrt{2}}\ \Omega$$

and the capacitance $C_1$ is

$$C_1 = \frac{1}{(2\pi \times 60) \times 10\cos(45°)} = \frac{\sqrt{2}}{2\pi \times 600}\ F$$

Therefore, the series impedance Z is

$$Z = \frac{10}{\sqrt{2}} - j\frac{10}{\sqrt{2}} = 10\angle -45°\ \Omega$$

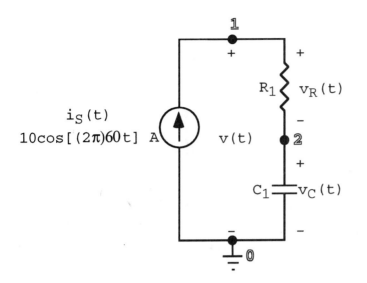

FIGURE P12.1

Determine the capacitance voltage at time zero if the circuit is in periodic steady-state operation. With this value of voltage as the capacitance initial condition voltage, do a PSpice transient simulation using the SIN function transient source specification for a time equal to three periods of the current source. Maintain six digits of precision. Select <t_Print> small enough to have at least twenty points per period. Have the simulation create a DAT file for Probe.

a) With Probe, plot the voltage v(t) as a function of the resistance voltage $v_R(t)$. From the resulting Lissajous pattern, evaluate the phase shift between the voltage v(t) and current i(t). Change the horizontal axis back to time. Now, plot both v(t) and $i_S(t)$ to see whether v(t) leads or lags $i_S(t)$. Use Probe's menu commands **Y_axis, add_aXis, Change_title, selecT_axis,** and **Add_trace** to display both variables with appropriate labeling for each axis and with each variable using the correct y-axis.

b) Remove the traces in Part a and plot the instantaneous power that the source delivers to the entire circuit. Use Probe's cursor to identify the maximum and minimum values of the instantaneous power. Average these values to obtain the average power. Subtract the average power from the maximum value to determine $V_m I_m/2$. Finally, divide this value into the average power to obtain the value of the power factor pf.

c) Superimpose the instantaneous power for the resistance and the capacitance on the plot of Part b. Add the average value of each of these waveforms to this plot, using Probe's avg(x) function. Also, plot the average power using

$$P = \frac{1}{2}V_m I_m \cos(\theta) = \frac{1}{2} \times 100 \times 10 \times \cos(45\pi/180)$$

to see Probe's running average of the instantaneous power converge to the average power.

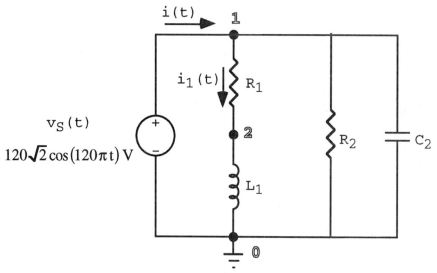

FIGURE P12.2

## 12.2 Power-Factor Improvement

(Chapter 12, Example 4)

For the circuit shown in Fig. P12.2, impedance $Z_1 = R_1 + j\omega L_1$ absorbs 1 kW with a lagging power factor of 0.9, and admittance $Y_2 = 1/R_2 + j\omega C_2$ absorbs 10 W with a leading power factor of 0.02. Determine impedance $Z_1$, admittance $Y_2$, and calculate the values of $L_1$ and $C_2$. Using PSpice, simulate the circuit with these element values. Measure the magnitude and phase of the current $i_1(t)$ and the total current $i(t)$. Using these phasor currents, confirm that your value of the impedance of $Z_1$ is correct, and calculate the composite impedance $Z = Z_1 || (1/Y_2)$.

## 12.3 Thévenin Equivalent and Maximum Power Transfer

For the circuit shown in Fig. P12.3 and using PSpice:

    a) Find the real and imaginary parts of the Thévenin equivalent impedance $Z_T$ at 1 kHz, as seen by the load circuit.

    b) With load resistance $R_L$ equal to the real part of $Z_T$ from Part a, use the STEP command to vary the ratio of the load inductance $L_L$ to its optimum value $L_{Lopt}$ from -1.0 to 2.0 in steps of 0.2. Print in the output file the magnitude and phase of the output voltage at node 2 and the load current. Using these values, calculate and plot the load power as a function of $L_L/L_{Lopt}$. PSpice allows use of negative or zero inductance values. The implication of a negative inductance value is that the imaginary part of the load impedance becomes capacitive.

    c) Repeat Part b using a load resistance value both twice and half the optimum value.

From the results of Parts b and c you can plot the load power as a function of relative inductance value with load resistance as a parameter. These curves show the behavior shown in Fig. 12.18 of Cunningham and Stuller.

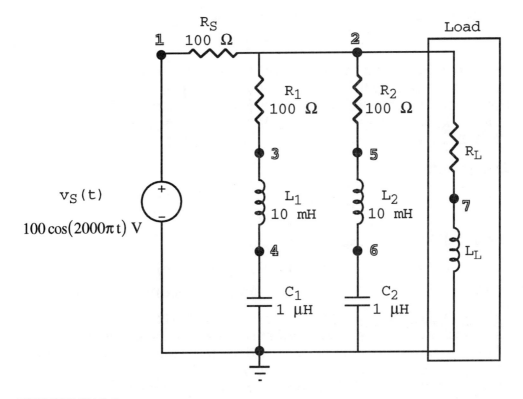

FIGURE P12.3

## 12.4 Power Triangle as a Function of Frequency

For the circuit shown in Fig. P12.4:

    a) Write a PSpice netlist to obtain a DAT file for Probe, using an rms value for the AC current source. Sweep the frequency from 10 Hz to 100 kHz.

    b) Show that the average power P that the source delivers to the load impedance is

$$P = \frac{V_{rms}^{2}}{R_1}$$

the reactive power Q is

$$Q = V_{rms}^{2}\left(\frac{1}{f} - \frac{f}{10^{6}}\right)$$

and the apparent power |S| is

$$|S| = \sqrt{P^{2} + Q^{2}}$$

c) With Probe, plot the average power P using

$$V(1) * V(1) / 100$$

Add to this plot the reactive power using

$$V(1) * V(1) * (1/\text{frequency} - \text{frequency}/1e6)$$

Finally, add the apparent power to the plot using

$$\text{sqrt}(\#1*\#1+\#2*\#2)$$

Here, #1 refers to the first trace, which is the average power P, and #2 refers to the second trace, which is the reactive power Q. Remember, Probe does complex arithmetic with AC expressions and displays the magnitude of the result. Therefore V(1)*V(1) is the same as VM(1)*VM(1).

d) Using the power factor equation

$$pf = \cos(\theta) = \frac{P}{|S|}$$

show that the Probe expression

$$1/\text{sqrt}(1+1e4*(1/\text{frequency}-\text{frequency}/1e6)$$
$$*(1/\text{frequency}-\text{frequency}/1e6))$$

defines the power factor. Use this expression to plot the power factor as a function of the frequency. What is the value of the frequency where the power factor is one? How does this frequency relate to the resonant frequency of the circuit? Why does the power factor approach zero at both low and high frequency values?

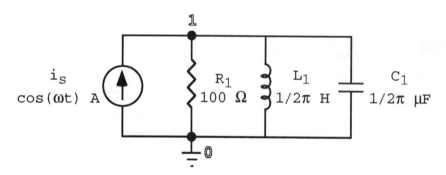

FIGURE P12.4

## Problems for Textbook Chapter 13

Chapter 13 of Cunningham and Stuller describes the effects of frequency on the magnitude and phase shift of circuit voltage and current responses. The common method for displaying magnitude and phase response functions is to plot the dB or log magnitude response and phase as functions of the logarithm of the frequency. These plots are known as Bode plots. Because the technique is graphical, we use Probe in all of the examples for this section.

## 13.1 RL Circuit Frequency Response

(Chapter 13, Section 1 - High-Pass Filter)
For the circuit shown in Fig. P13.1 use PSpice and Probe to plot the dB magnitude and phase of the transfer functions relating phasor resistance

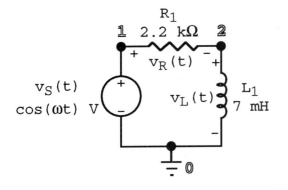

FIGURE P13.1

voltage $V_r$ and phasor inductance voltage $V_l$ to the source $V_s$ as a function of the frequency for frequency values ranging from 500 Hz to 5 MHz. Mark the magnitude curves with their half-power frequency values and the phase curves with their ±45° phase-shift values. Use the cursor function to find the half-power frequencies. Does the value that you measure agree with $R_1/(2\pi L_1)$?

## 13.2 RC Circuit Frequency Response

(Chapter 13, Section 1 - Low-Pass Filter)

For the circuit shown in Fig. P13.2, use PSpice and Probe to plot the dB magnitude and phase of the transfer functions relating phasor resistance voltage $V_r$ and phasor capacitance voltage $V_c$ to the source $V_s$ as a function of the frequency for frequency values ranging from 10 mHz to 100 Hz. Mark the magnitude curves with their half-power values and the phase curves with their 45° phase-shift values. Use Probe's cursor to find the half-power frequency. Does the value that you measure agree with $1/(2\pi R_1 C_1)$?

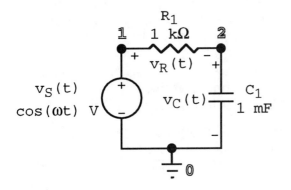

FIGURE P13.2

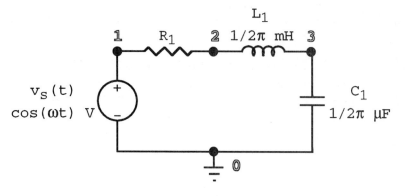

FIGURE P13.3

## 13.3 Resonance in a Series RLC Circuit

(Chapter 13, Section 2)

Simulate the circuit shown in Fig. P13.3 with PSpice and create a DAT file for Probe. Let $R_1$ have values equal to 0.1, 0.2, and 0.5 $\Omega$. Vary the frequency from 10 kHz to 100 MHz with 50 points per decade.

a) Using Probe, plot the phasor voltages across the resistance, inductance, and capacitance for $R_1$ equal to 0.5 $\Omega$ as a function of log(f). Your result shows that the resistance voltage maximizes at a center frequency $f_o$. The inductance voltage maximizes above $f_o$ and the capacitance voltage maximizes below $f_o$. The product of these two frequency maxima equals $f_o$. At low frequencies the resistance voltage goes to zero linearly, and the inductance voltage goes to zero with the square of the frequency. At high frequencies the resistance voltage goes to zero inversely with frequency, and the capacitance voltage goes to zero inversely with the square of the frequency. At low frequency the capacitance voltage goes to one. At high frequency the inductance voltage goes to one. Measure the half-power frequency values where the resistance voltage equals $1/\sqrt{2}$ V.

b) Using Probe, plot the phase of the phasor voltages across the resistance, inductance, and capacitance for $R_1$ equal to 0.5 $\Omega$ as a function of log(f). Your results here show that the phase of all three voltages has the same variation with frequency, but the inductance voltage leads and the capacitance voltage lags the resistance voltage by 90°. Measure thefrequency values where the phase shift equals ±45° for the resistance voltage. You will find that these frequency values are the same as for the half-power frequencies in Part a.

c) Using Probe plot the resistance voltage dB magnitude vs. log(f) for all three values of resistance. Next, plot the phase vs. log(f). Repeat for the inductance voltage and capacitance voltage dB magnitude and phase.

## 13.4 Resonance in a Parallel RLC Circuit

(Chapter 13, Section 5 - Parallel RLC Circuit)

Repeat Problem 13.3 for a parallel RLC circuit having resistance $R_1$ that assumes values of 1, 2, and 5 kΩ. The capacitance equals $1/\pi$ nF and the inductance value is $250/\pi$ μH. The AC source is a current source having a value of 1 mA and 0° phase. Plot the currents through the resistance, inductance, and capacitance.

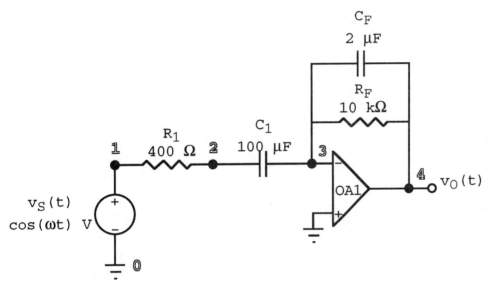

FIGURE P13.5

## 13.5 Op-Amp RC Filter

(Chapter 13, Example 7)

Simulate the circuit shown in Fig. P13.5 using PSpice. Include the PROBE command in your netlist. Plot the Bode magnitude and phase plots using Probe. Print hard copies of these plots and superimpose straight-line Bode magnitude and phase curves on these plots to see how effectively the straight lines describe the actual curves.

## 13.6 Twin-Tee Notch Filter Circuit

For the twin-tee notch-filter circuit shown in Fig. P13.6

a) Write the node equations and solve these using Cramer's rule to show that the transfer function giving the ratio of the output voltage $V_3$ at node 3 to the source voltage $V_s$ is

$$F(s) = \frac{s^2 + \omega_o^2}{s^2 + (\omega_o/Q)s + \omega_o^2}$$

Find how the center frequency $\omega_o$ and quality factor Q depend on the values of R and C.

b) Identify the location of the poles and zeros of this function.

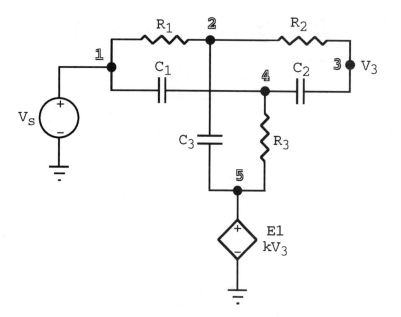

Fig. P13.6 High-Q Twin-Tee Circuit ($R_1 = R_2 = 2R_3 = R = 10 \text{ k}\Omega$, $C_1 = C_2 = C_3/2 = C = 100/2\pi$ nF)

c) Letting $s = j\omega$ and normalizing the frequency by $\omega_o$, write expressions for the magnitude and angle of the transfer function when $s = j\omega$. Determine the frequencies where the magnitude equals $1/\sqrt{2}$ times the low- and high-frequency values. Show that these frequencies are the same as the frequencies where the phase shift equals $\pm45°$.

d) Simulate the circuit with the parameter values shown in Fig. P13.6 and using PSpice. Vary the frequency from 100 Hz to 10 kHz, using 20 points per decade. Step the E-source parameter k so that Q equals 0.25, 0.5, 1.0, and 2.0. Using Probe, plot the magnitude and the phase of voltage $V_3$ at node 3 as a function of the frequency and with Q as a parameter.

e) Do a 50-point Monte Carlo simulation with a $\pm10\%$ Gaussian device variation on all resistances and capacitances. Vary the frequency from 100 Hz to 10 kHz, using 20 points per decade, but set the Q value at 2.0. Write goal functions to measure the center frequency and half-power bandwidth, storing these functions in an appropriately named goal-function file. Using Probe, obtain histograms showing the center frequency and the half-power bandwidth statistical variations.

## Problems for Textbook Chapter 14

This section provides problems that involve Laplace transform evaluation of transients in electrical circuits. The analog behavioral modeling feature of E- and G-type controlled sources includes a LAPLACE option that lets you model the output of these two source types in terms of a Laplace transform. Because the computation procedure is slow, we do not recommend use of this feature.

Nevertheless, you can check computations for some simple transforms with the LAPLACE option.

## 14.1 RC Circuit with Equal Source and Circuit Frequency

Simulate the circuit shown in Fig. P14.1 using the LAPLACE E-source option so that

$$V_{E1}(s) = \frac{s}{s+\alpha} V_{s1}(s)$$

where $\alpha$ equals 1 krad/s. Use the TRAN command to obtain the circuit transient response when $v_{S1}(t)$ is a 1-V voltage step. Also, give voltage source $v_{S1}(t)$ a 1-V AC magnitude with 0° of phase shift and use the TRAN command, sweeping the frequency with 20 steps per decade from 1 Hz to 10 kHz.

> a) Use Probe to display the transient waveforms for $v_{E1}(t)$, $v_{R1}(t)$, and $v_{C1}(t)$. Prove that
>
> $$v_{E1}(t) = V_{E1} e^{-\alpha t} u(t)$$
> $$v_{R1}(t) = V_{E1}(1 - \alpha t) e^{-\alpha t} u(t)$$
>
> and
>
> $$v_{C1}(t) = \alpha t e^{-\alpha t} u(t)$$
>
> where $V_{E1}$ equals 1 V and $\alpha$ equals 1 krad/s. Superimpose these solutions on the Probe plot to confirm the calculations.

> b) Use Probe to display the frequency response for the phasor dB magnitudes of $v_{E1}(t)$, $v_{R1}(t)$, and $v_{C1}(t)$. Show that
>
> $$\left|V_r(j\omega)\right| = \frac{V_{s1m}}{1 + \left(\dfrac{\alpha}{\omega}\right)^2}$$
>
> $$\left|V_{e1}(j\omega)\right| = \sqrt{\left|V_r(j\omega)\right|}$$

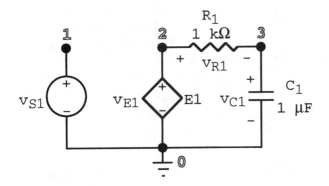

FIGURE P14.1

and

$$\left|V_c(j\omega)\right| = \frac{\alpha}{\omega}\left|V_r(j\omega)\right|$$

where $V_{s1m}$ equals 1 V and $\alpha$ equals 1 krad/s. Use these to plot the theoretical dB magnitudes on your Probe plot to confirm the calculations.

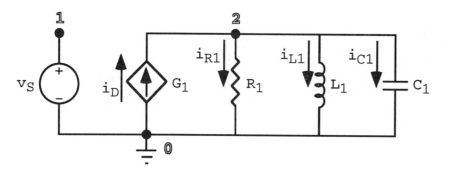

FIGURE P14.2

## 14.2 Parallel RLC Circuit with Equal Source and Circuit Frequency

Voltage $v_S(t)$ in the circuit of Fig. P14.2 has a transient 1-V unit step component. Set the transconductance of the G1 source so that the Laplace transform of dependent source $i_D(t)$ is

$$I_D(s) = \frac{5\omega_n}{(s+\alpha)^2 + \omega_n^2} \quad mA - s$$

when $v_S(t)$ is a 1-V unit step. The damping constant $\alpha$ equals $2\pi/10$ MNep/s and the damped natural frequency $\omega_n$ equals $2\pi\sqrt{99}/10$ Mrad/s. The undamped natural frequency $\omega_o$

$$\omega_o = \sqrt{\alpha^2 + \omega_o^2}$$

equals $2\pi$ Mrad/s. Design the circuit parameters $R_1$, $L_1$ and $C_1$ so that the impedance at the undamped natural frequency equals 5 k$\Omega$ and the quality factor $Q = \omega_o R_1 C_1$ is 5. Use PSpice to simulate the circuit, giving the voltage source $v_S(t)$ a $1\angle 0°$ V AC component in addition to the 1-V unit step transient component. Use the AC, TRAN, and PROBE commands so that you can use Probe to view the AC and transient response of $i_D(t)$, $i_{R1}(t)$, $i_{L1}(t)$, and $i_{C1}(t)$. With Probe, superimpose all four variables to see how they compare with each other. Plot each one separately and superimpose on each plot a mathematical solution that you obtain using phasor or Laplace domain analysis.

## 14.3 Three-Pole Butterworth Transient

The circuit shown in Fig. P14.3 is a three-pole Butterworth filter circuit.

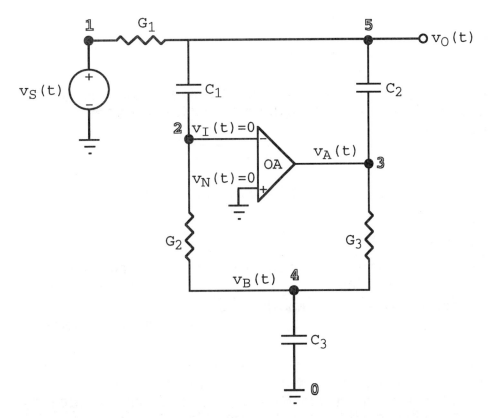

FIGURE P14.3 Three-Pole Butterworth Filter ($R_1 = 1/G_1 = R_2 = 1/G_2 = R_3 = 1/G_3 = 1/G = 1$ k$\Omega$, $C_1 = C_2 = C_3 = C = 1$ $\mu$F)

a) Simulate this circuit when $v_S(t)$ is a 1-V unit step. Use Probe to plot the output voltage waveform $v_O(t)$. Determine the Laplace transform $V_O(s)$ for the circuit, take the inverse transform to write $v_O(t)$, and superimpose this solution on the plot.

b) Repeat when $v_S(t)$ is a 1 V-ms unit impulse.

c) Repeat when $v_S(t)$ is

$$v_s(t) = e^{-t/2} \cos(\sqrt{3}t/2) \text{ V}$$

where t is in ms.

## Problems for Textbook Chapter 15

Problems in this section show how to use PSpice to determine Fourier series representations of periodic waveforms.

### 15.1 Fourier Series of a Periodic Square Wave

(Chapter 15, Example 1)

Simulate the circuit shown in Fig. P15.1a where $v_S(t)$ is the 5-V square wave shown in Fig. P15.1b. Use the TRAN command with the print-step time <t_Print> equal to 2 $\mu$s and the simulation time <t_Final> equal to 1 ms. Use

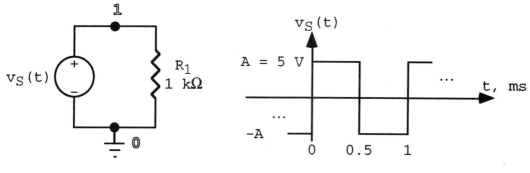

a) Circuit.                                    b) Square-Wave Waveform

FIGURE P15.1

the FOUR command to obtain the Fourier components of the voltage $v_S(t)$. Simulate the source voltage using

a) A PWL function. Use the breakpoints (0, 0), (1u, 5), (0.499m, 5), (0.501m, -5), (0.999m, -5), and (1m, 0) to approximate the square-wave function.

b) A PULSE function. Set the starting level $V_1$ equal to -5 V, the pulse level $V_2$ equal to 5 V, the delay time $t_d$ equal to zero, both the rise time $t_r$ and fall time $t_f$ equal to 2 μs, the pulse duration $t_p$ equal to 0.5 ms, and the period $t_{per}$ equal to 1 ms. Compare the Fourier component magnitudes and phases from the simulation with values that you expect from

$$v_s(t) = \sum_{n=1,3,5,\ldots}^{\infty} \frac{4A}{n\pi} \sin(2\pi n f_i t)$$

Notice that the phases in the PSpice table of Fourier values are consistent with a sine function rather than a cosine function. Are the even harmonics zero? Notice that the even harmonics for the PULSE approximation of the square wave are not as small as for the PWL approximation. The PULSE approximation of the square wave in Part b is not exactly half-wave symmetric. Change a parameter of the PULSE function so that it becomes half-wave symmetric and repeat the simulation. Now both simulations have small-valued even harmonics of about the same size.

c) Change the voltage parameters for the PULSE function of Part b so that the voltage waveform $v_S(t)$ has a maximum value of 10 V and a minimum value of 0 V and repeat the simulation of Part b. Compare the SPICE Fourier table here with the table from Part b.

## 15.2 Fourier Series of a Pulse with Variable Duty Cycle

(Chapter 15, Example 3)
Simulate the circuit of Fig. P15.1 where voltage source $v_S(t)$ is the variable duty-cycle pulse shown in Fig. P15.2. Let the duty cycle parameter τ have

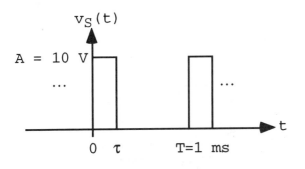

FIGURE P15.2

values of 0.125, 0.2, and 0.25 μs using the STEP command. Approximate the source waveform using the PULSE transient source specification with a 0.5-μs rise- and fall time and a pulse duration equal to the pulse duration τ less 1 μs. Use the TRAN command with the print-step time <t_Print> equal to 0.5 μs, so that Probe can accurately display the rise- and fall times. Set the simulation time <t_Final> equal to the 1-ms period. Use the FOUR and PROBE commands so that you can look at Fourier series data in the output file, and use Probe to plot the Fourier components.

Examine your output file to see that the Fourier coefficient amplitudes agree with

$$A_n = 2A \frac{\tau}{T} \left| \frac{\sin(\pi n f_1 \tau)}{\pi n f_1 \tau} \right|$$

Notice that the magnitudes go to zero when the index n is a multiple of T/τ. Also, notice that the phase-shift values differ from the 90° values that you can expect when the source waveform is the even function of time shown in Fig. P15.9 of Cunningham and Stuller. Remember, SPICE presents Fourier series in sine-function form, not the cosine-function form. Since SPICE does not allow time values less than zero, the waveform in this PSpice problem has a positive time translation of τ/2. However, the time-shift property P3 in Table 15.2 of Cunningham and Stuller shows that time translation of a periodic waveform by τ/2 does not change the amplitude of the harmonics but decreases the phase shift by 180nτ/T degrees. As a result, the phase-shift values that appear in the SPICE Fourier series tables are 90 - 180nτ/T in degrees.

Use Probe's **Add_trace** menu command to select and plot the source voltage $v_S(t)$ for each value of pulse duration τ. You can use the notation V(1)@1 to identify the source voltage for the first case of the STEP command. Select **Fourier** from the X_axis menu to display the Fourier transform of the waveform. Use **Set_range** from the X_axis menu to set the frequency range to extend from 0 to 40 kHz to see the sin(x)/x form of the Fourier coefficients. Reduce the frequency range to extend from 0 to 10 kHz, and use the cursor to read the Fourier coefficient amplitudes at each harmonic. Plot the phase shift for each case.

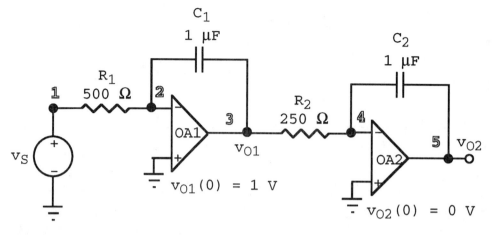

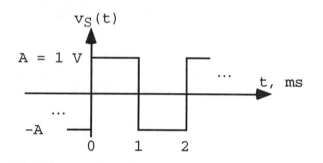

a) Circuit.

b) Square-Wave Waveform.

FIGURE P15.3

## 15.3 Square, Triangle, and Parabolic Fourier Series

For the circuit shown in Fig. P15.3, use PSpice to compute and display the Fourier coefficients of the source voltage $v_S(t)$, the output voltage $v_{O1}(t)$, and the output voltage $v_{O2}(t)$. Approximate the source voltage $v_S(t)$ using a PWL transient specification with breakpoints (0, 0), (10u, 1), (0.990m, 1), (1.01m, -1), (1.990m, -1), and (2m, 0).

## 15.4 Full-Wave Rectified Sinusoidal Source with LC Filter

Using PSpice and Probe, plot the waveforms of the source voltage $v_S(t)$ and load voltage $v_O(t)$ and the current $i_L(t)$ for the LC-filter circuit shown in Fig. P15.4. Also, plot the running average of the instantaneous power $P_S$ that the source supplies and the instantaneous power $P_L$ that the load receives. The source for this circuit is a full-wave rectified sinusoidal waveform. Display two cycles of the 120-Hz fundamental period of the full-wave rectified source after the circuit arrives at steady-state operation. Use an E-type source to represent the full-wave rectified source $v_S(t)$. With PSpice, evaluate the Fourier series coefficients of the voltages $v_S(t)$ and $v_O(t)$ and current $i_L(t)$ for the first five harmonics. Select <t_Print> so that there are 50 points for each 120-Hz period.

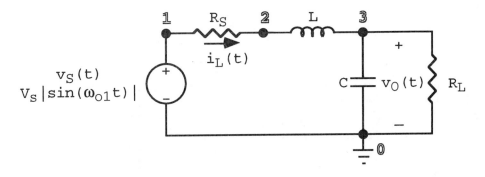

FIGURE P15.4 ($V_s = 440\sqrt{2}$ V, $f_{o1} = 60$ Hz, $R_S = 50\ \Omega$, $R_L = 4\ k\Omega$, $L = 8$ H, $C = 33\ \mu F$)

Calculate the Fourier series coefficients of voltages $v_S(t)$ and $v_O(t)$ and of current $i_L(t)$ for the DC term and the first five harmonics. Calculate the average powers $P_S$ and $P_L$. Compare your calculations with the Probe observations.

## Problems for Textbook Chapter 16

The problems in this section show how to simulate two-port circuits to determine their terminal parameters.

### 16.1 Z-Parameter Circuit

(Chapter 16, Example 1)

Simulate the circuit shown in Fig. P16.1 to determine the open-circuit voltages $V_{1oc}$ at node 1 and $V_{2oc}$ at node 2 with respect to the reference node. Also, determine the two-port impedances $Z_{11}$, $Z_{12}$, $Z_{21}$, and $Z_{22}$.

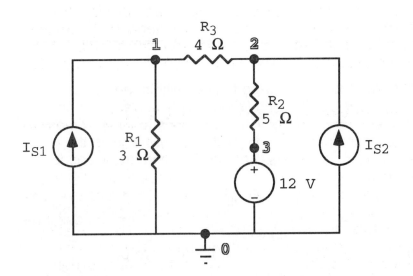

FIGURE P16.1

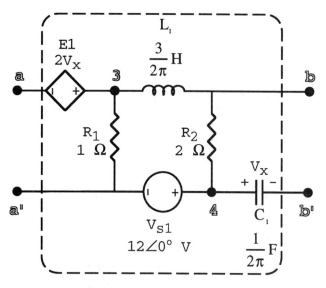

a) Two-Port Circuit.

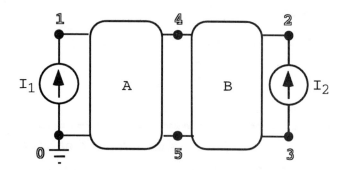

b) Cascade of Two Two-Port Circuits.

FIGURE P16.2

## 16.2 Z Parameters of Phasor Circuit

(Chapter 16, Example 5)

a) Use a subcircuit to describe the two-port circuit within the dotted outline of Fig. 16.2a for PSpice simulation. Apply current sources $I_1$ and $I_2$ at the input and output ports. Run the simulation three times with appropriate combinations of $I_1$, $I_2$, and $V_{s1}$ to determine the open-circuit voltages $V_{1oc}$ and $V_{2oc}$ and the two-port impedances $Z_{11}$, $Z_{12}$, $Z_{21}$, and $Z_{22}$.

b) Use PSpice to determine the open-circuit voltages $V_{1ocT}$ and $V_{2ocT}$ and the two-port impedances $Z_{11T}$, $Z_{12T}$, $Z_{21T}$, and $Z_{22T}$ that describe the circuit of Fig. P16.2b, which consists of a cascade connection of two of the two-port circuits of Part a. Show that the two-port parameters for this cascade connection are

$$V_{1ocT} = V_{1ocA} + \frac{Z_{12A}}{Z_{22A} + Z_{11B}}\left(V_{1ocB} - V_{2ocA}\right) = V_{1oc} + \frac{Z_{12}}{Z_{11} + Z_{22}}\left(V_{1oc} - V_{2oc}\right)$$

$$V_{2ocT} = V_{2ocB} + \frac{Z_{21B}}{Z_{22A} + Z_{11B}}\left(V_{2ocA} - V_{1ocB}\right) = V_{2oc} + \frac{Z_{21}}{Z_{11} + Z_{22}}\left(V_{2oc} - V_{1oc}\right)$$

$$Z_{11T} = Z_{11A} - \frac{Z_{12A}Z_{21A}}{Z_{22A} + Z_{11B}} = Z_{11} - \frac{Z_{12}Z_{21}}{Z_{11} + Z_{22}}$$

$$Z_{12T} = \frac{Z_{12A}Z_{12B}}{Z_{22A} + Z_{11B}} = \frac{Z_{12}^{\,2}}{Z_{11} + Z_{22}}$$

$$Z_{21T} = \frac{Z_{21A}Z_{@1B}}{Z_{22A} + Z_{11B}} = \frac{Z_{21}^{\,2}}{Z_{11} + Z_{22}}$$

$$Z_{22T} = Z_{22A} - \frac{Z_{12B}Z_{21B}}{Z_{22A} + Z_{11B}} = Z_{22} - \frac{Z_{12}Z_{21}}{Z_{11} + Z_{22}}$$

Use these equations to calculate the two-port values for the cascade connection and compare the calculations with a PSpice simulation.

## Problems for Textbook Chapter 17

The problems in this section show how to simulate inductance circuits that have mutual coupling.

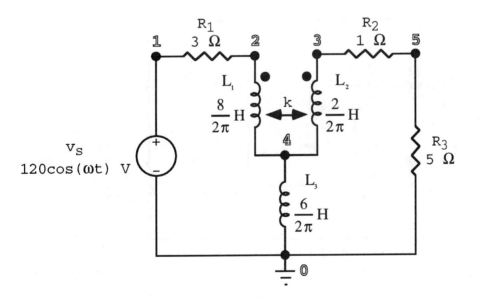

FIGURE P17.1

## 17.1 Impedance of Mutual Inductance Circuit

(Chapter 17, Problem 6)

Use PSpice to simulate the circuit shown in Fig. P17.1 and plot the magnitude and phase of the impedance $Z_{in} = V_s/I_{r1}$ using Probe. Vary the frequency from 0.01 to 100 Hz using 20 points per decade and step the mutual coupling from -1 to 1 with increments of 0.25. Use **Y_axis** and **Set_range** to set the y axis range to 0, 500 $\Omega$ for the magnitude plot. Reduce the range further to see the low-frequency limit of the magnitude plot.

## 17.2 Circuit with Ideal Transformer

(Chapter 17, Example 3)

Using E- and F-type controlled sources to model the ideal transformer in the circuit shown in Fig. P17.2, use PSpice to determine the phasor currents $I_1$, $I_2$, and $I_3$ and the phasor voltages $V_2$ and $V_3$.

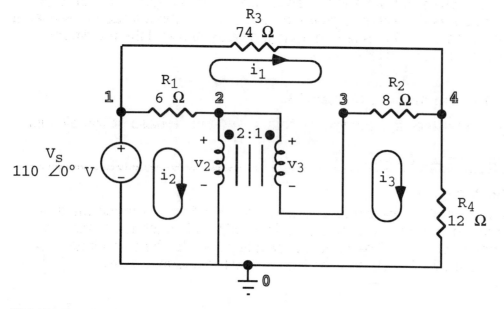

FIGURE P17.2

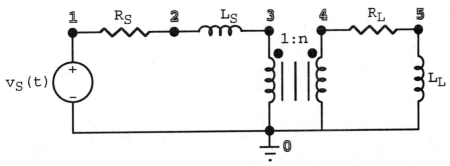

FIGURE P17.3 $R_S = 1\ \Omega$, $L_S = 4/(2\pi\times60)$ H,  $R_L = 80\ \Omega$, $L_L = (1/2\pi)$ H,  and
$v_S(t) = 440\sqrt{2}\ \cos(\omega t)$

## 17.3 Coefficient of Coupling for Maximum Power

a) Using PSpice, simulate the ideal-transformer circuit shown in
Fig. P17.3 for values of frequency ranging from 6 to 600 Hz using 20
points per decade. Change the ideal transformer turns ratio n from 2
to 8 in steps of 0.2 using the STEP command. Include the PROBE
command in the netlist.

b) Run Probe using the DAT file PR17_3s.DAT that PSpice produces
when the solution netlist PR17_3s.CIR is the netlist file. Define a
macro function to calculate the average power. Using this macro,
plotthe average power that the input port of the ideal transformer
absorbs as a function of the frequency and with turns ratio n as
parameter.

c) Use Probe's performance analysis feature and a goal function that
measures the average power at 60 Hz to plot the average power as a
function of the turns ratio n. Find the value of n and the maximum
power using Probe's cursor.

## Problems for Textbook Chapter 18

The problem in this section shows how to simulate three-phase power circuits.

## 18.1 Three-Phase Circuit

(Chapter 18, Example 4)
Use PSpice to simulate the circuit shown in Fig. P18.1 to determine the line
currents from the source to the loads, the line currents in each load, and the
delta load phase currents. The sources operate at 60 Hz, the wye circuit
impedances are 80 - j10 $\Omega$, and the delta-circuit impedances are
200 + j200 $\Omega$.

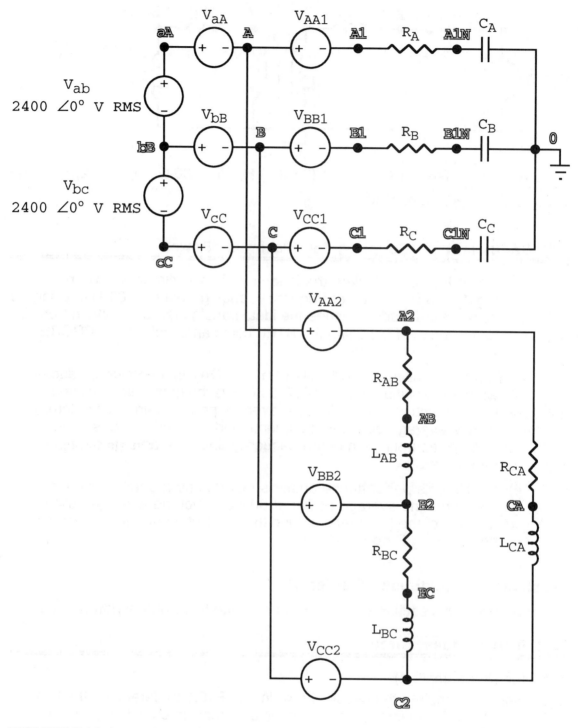

FIGURE P18.1

# Chapter 4 Solutions to Selected Problems

This chapter provides solutions for the following simulation problems from Chapter 3:

- 9.4 RLC Circuit Switching Transient
- 13.6 Twin-Tee Notch Filter Circuit
- 14.3 Three-Pole Butterworth Transient
- 15.4 Full-Wave Rectified Sinusoidal Source with LC Filter
- 17.3 Coefficient of Coupling for Maximum Power

These five sample solutions show how to simulate transient and phasor circuit problems using cursor control and measurement, performance analysis, Fourier series measurement, and Monte Carlo statistical analysis. The solutions demonstrate that effective simulation depends on your ability to use circuit analysis or insight to establish appropriate time values for transient simulations and frequency values for AC simulations.

## 9.4 RLC Circuit Switching Transient

The netlist for this problem appears in Fig. 4.1. The MODEL statement for switch S1 sets the on resistance RON to 1 mΩ and the off resistance ROFF to 10 MΩ. With these choices, the resistance between nodes 1 and 3 has a value of 50.001 Ω instead of 50 Ω with the switch on and 199.998 Ω instead of 200 Ω with the switch off. The ratio of $R_{OFF}/R_{ON}$ is $10^{10}$, which is less than the maximum usable ratio of $10^{12}$.

The switch control voltage $v_C$ is a PULSE waveform that starts high at 1 V, drops quickly to 0 V, remains low until t = 1 s, then quickly returns high, and remains high until t = 2 s. To choose the time duration of the switch control-voltage transitions, we need to know the damping coefficients of the circuit. When the switch opens, the circuit is a series RLC circuit with resistance $R_1 + R_2$, so the characteristic equation is

$$s^2 + \frac{R_1 + R_2}{L_1}s + \frac{1}{L_1 C_1} = 0$$

$$s^2 + 40s + 144 = 0 \qquad\qquad s_{1,2} = -4\,\text{Nep/s}, -36\,\text{Nep/s}$$

For the switch to appear to open instantaneously, the control voltage has to change from 1 to 0 V in a time much smaller than either 1/4 or 1/36 s. Therefore, we choose the rise time of the PULSE waveform to be 1 ms. For the second time interval, when the switch closes again, the series resistance reduces to $R_2$, and the characteristic equation becomes

$$s^2 + \frac{R_2}{L_1}s + \frac{1}{L_1 C_1} = 0$$

$$s^2 + 10s + 144 = 0 \qquad\qquad s_{1,2} = -5 \pm j\sqrt{119}\,\text{rad/s}$$

Now, closing the switch in a time much smaller than 1/5 s appears to the circuit as an instantaneous change. The netlist sets the fall time to be 1 ms. The

```
Problem 9.4 - RLC Circuit Switching Transient
*
*       See Chapter 9, Problem 20.
*
Is1    0     1       16
Vc     4     0       PULSE(1, 0, 0, 1m, 1m, 999m, 2)
R1     1     2       150
R2     2     3       50
L1     3     0       5
C1     1     0       {1/720}
S1     1     2       4       0       Smod
.MODEL Smod VSWITCH       (RON=1m, ROFF=10Meg)
.TRAN 5m     2       0       5m
.PROBE
.END
```

FIGURE 4.1 PSpice Netlist for Problem 9.4

control voltage turns the switch off for all but the first 1 ms of the first 1-s interval and turns the switch back on for all but 1 ms of the second 1-s interval.

To have five output points occur during one time constant of the fast time-constant transient during the first 1-s interval, requires that <t_Print> be about 5 ms. The slow transient runs to completion in about five time constants, which equals 1.25 s. Consequently, steady-state operation does not exist when the switch closes at 1 s. The transient that occurs when the switch closes again has a damping factor of 5 Nep/s. Having 5 points occur within 1/5 s requires that <t_Print> equal 40 ms, so the selection of 5 ms above is acceptable here. Since five time constants equals 1 s, letting <t_Final> be 2 s allows the circuit to recover to steady state. The oscillation period equals $2\pi/\sqrt{119} \approx 576$ ms, so having <t_Print> equal to 5 ms gives more than 100 points per period. You can probably use the 40-ms default value of <t_Ceil> for this simulation and have PSpice accurately calculate the transient with the switch on. However, the netlist sets <t_Ceil> equal to 5 ms.

Figures 4.2 and 4.3 show the Probe plots of v(t) and i(t) that result using the netlist of Fig. 4.1. The overdamped behavior during the first interval and underdamped behavior for the second interval are apparent. If you run this simulation and do the Probe plot, measure the voltage and current with the cursor at t = 0 and t = 1 s to see that their values at these times agree with the values that the following calculations use.

Now, calculate the voltage v(t) and current i(t) to superimpose on the PSpice waveforms for comparison. The equations for the voltage and current are

$$v(t) = 3200 + V_1 e^{-4t} + V_2 e^{-36t}$$

and

$$i(t) = 16 + I_1 e^{-4t} + I_2 e^{-36t}$$

because the steady-state DC voltage v(∞) = (R₁ + R₂)Is = 3200 V and the steady-state DC current i(∞) = is = 16 A. The initial conditions are

$$i(0+) = i(0-) = 16\,A$$

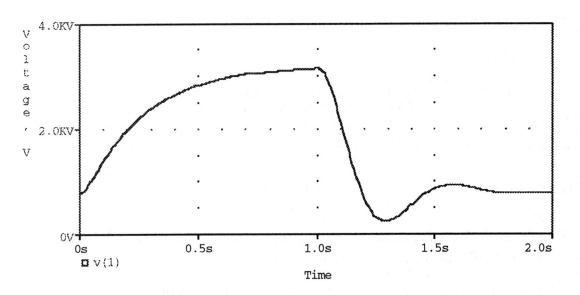

FIGURE 4.2 Voltage  for  $0 \leq t \leq 2$ s

$$v(0+) = v(0-) = R_2 i(0+) = 50 \times 16 = 800 \text{ V}$$

$$i'(0+) = \frac{v_L(0+)}{L_1} = \frac{v(0+) - (R_1 + R_2)i(0+)}{L_1} = \frac{800 - 200 \times 16}{5} = -480 \text{ A/s}$$

$$v'(0+) = \frac{i_c(0+)}{C_1} = \frac{i_s - i(0+)}{C_1} = \frac{16 - 16}{5} = 0 \text{ V/s}$$

Substituting these into the current and voltage equations gives

$$v(t) = \left(3200 - 2700e^{-4t} + 300e^{-36t}\right)u(t) \text{ V}$$

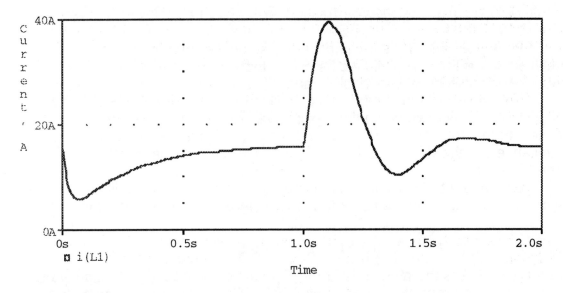

FIGURE 4.3 Current  for  $0 \leq t \leq 2$ s

and

$$i(t) = \left(16 - 15e^{-4t} + 15e^{-36t}\right)u(t) \text{ A}$$

Calculation of the voltage and current values at 1 s gives the initial values for the second interval. These are v(1) = 3150.55 V and i(1) = 15.7253 A. Since $v(\infty) = R_2 i_{S1} = 800$ V and $i(\infty) = 16$ A for the second 1-s interval, the voltage and currents are

$$v(t) = \left\{800 + V_1 e^{-5(t-1)} \cos\left[\sqrt{119}(t-1) + \phi_v\right]\right\} \text{ V}$$

and

$$i(t) = \left\{\left(16 + I e^{-5(t-1)} \cos\left[\sqrt{119}(t-1) + \phi_I\right]\right)\right\} \text{ A}$$

At t = 1 s the initial conditions are

$$i(1+) = i(1-) = 15.7253 \text{ A}$$

$$v(1+) = v(1-) = 3150.55 \text{ V}$$

$$i'(1+) = \frac{v(1+)}{L_1} = \frac{v(1+) - R_2 i(1+)}{L_1} = \frac{3150.55 - 200 \times 15.7253}{5} = 61.098 \text{ A/s}$$

$$v'(1+) = \frac{i_c(1+)}{C_1} = \frac{I_s - i(1+)}{C_1} = \frac{16 - 15.7253}{1/720} = 197.784 \text{ V/s}$$

Substituting these values into the equations for the current and voltage gives

$$v(t) = \left\{800 + 2593.3e^{-5(t-1)} \cos\left[\sqrt{119}(t-1) - 24.9885°\right]\right\}u(t) \text{ V}$$

and

$$i(t) = \left\{\left(16 + 43.2217e^{-5(t-1)} \cos\left[\sqrt{119}(t-1) - 90.3642°\right]\right)\right\}u(t) \text{ A}$$

Figures 4.4 and 4.5 superimpose the solutions onto the PSpice simulation waveforms. The slight differences that exist are due to non-instantaneous switching, imperfect on and off resistance, PSpice simulation error, or round-off error in the calculations. Because the switching event after t = 0 occurs at a time when the first 1-s interval transient has not come to completion, the solution needs to follow both the inductance current and the capacitance voltage during the first interval to obtain the initial condition for the second interval.

To measure the slow and fast damping factors and the amplitudes of the inductance current components for the first 1-s interval, notice that i(t) is given by

$$i(t) = I_{S1} + I_{FA}e^{-\alpha_{FA}t} + I_{SL}e^{-\alpha_{SL}t}$$

where $I_{S1} = 16$ A is the amplitude of the DC current source $i_{S1}$. Continuity of the inductance current at t = 0 requires that i(0+) = i(0-) = $I_{S1}$, so the slow

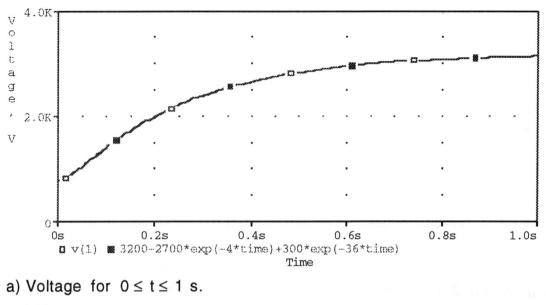

a) Voltage for $0 \le t \le 1$ s.

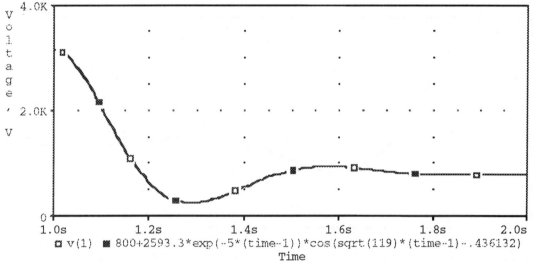

b) Voltage for $1 \le t \le 2$ s.

FIGURE 4.4 Comparison of PSpice and Voltage Calculation

transient amplitude $I_{SL}$ equals the negative of the fast-transient amplitude $I_{FA}$. Then

$$i(t) = I_{S1} + I_{FA}e^{-\alpha_{FA}t} - I_{FA}e^{-\alpha_{SL}t}$$

To measure the fast damping constant $\alpha_{FA}$, the slow damping constant $\alpha_{SL}$, and the amplitude $I_{FA}$, notice that the minimum occurs for

$$\frac{\ln(\alpha_{FA}/\alpha_{SL})}{\alpha_{FA}/\alpha_{SL} - 1} = \alpha_{SL}t_{min}$$

where $t_{min}$ is the time of the minimum. The fast transient becomes negligible at and beyond time $t_{min}$. Using a second point at time $t_2 = 2t_{min}$, where $i(t_2) = I_2$, and a third point at $t_3 = 3t_{min}$, where $i(t_3) = I_3$ gives

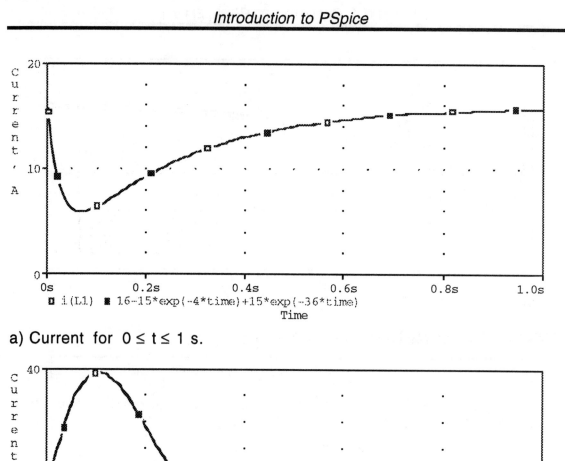

a) Current for $0 \le t \le 1$ s.

b) Current for $1 \le t \le 2$ s.

FIGURE 4.5 Comparison of PSpice and Current Calculation

$$\alpha_{SL} = \frac{\ln\left[(I_{S1} - I_2)/(I_{S1} - I_3)\right]}{t_3 - t_2}$$

From the third point

$$I_{FA} = -I_{SL} = \frac{I_{S1} - I_3}{e^{-\alpha_{SL}t_3}}$$

Figure 4.6 shows Probe's plot of i(t) for time between 0 and 1 s. Using the **Min** command on the cursor menu to move Cursor 1 to the minimum, gives $i(t_{min}) = $ 5.8613 A and $t_{min} = 69.128$ ms. Now move cursor 1 to $t_3 = 3t_{min} = $ 207.38 ms using the search command

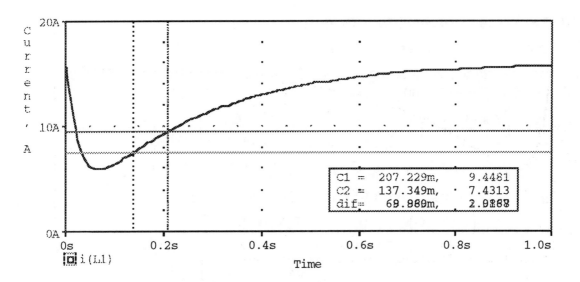

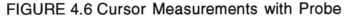

FIGURE 4.6 Cursor Measurements with Probe

search for xval (207.38m)

Next, holding the  <Shift> key while selecting **Search_Commands**, move the second cursor to time $t_2 = 2t_m$ = 138.26 ms using

search for xval (138.26m)

Figure 4.6 shows the values that these cursor movements give. Using these values

$$\alpha_{SL} = \frac{\ln[(16 - 7.4594)/(16 - 9.4520)]}{0.20738 - 0.13826} = 3.844 \ Nep/s$$

The fast-mode transient amplitude is

$$I_{FA} = -I_{SL} = \frac{16 - 9.4520}{e^{-3.844 \times 0.20738}} = 14.5 \ A$$

Finding the value of the ratio of $\alpha_{FA}$ to $\alpha_{SL}$, either using a programmable calculator or by iteration, gives $\alpha_{FA} = 9.454\alpha_{SL} = 9.454 \times 3.844 = 36.3$ Nep/s. Of course, the theoretical values are $I_{FA} = 15$ A, $\alpha_{FA} = 36$ Nep/s, and $\alpha_{SL} = 4$ Nep/s. The difference between the theoretical values and the Probe simulation values is due to our assumption that the fast-mode transient is negligible at $2t_{min}$ and $3t_{min}$, the 1-ms delay in the switch opening, the on and off resistance values not being 0 and ∞, and numerical inaccuracy due to working with 5-ms print-step and step-ceiling values. A curve-fit program gives

$$i(t) = 16.001 - 15.091e^{-4.0106t} + 15.091e^{-35.211t}$$

as the best fit to the PSpice data for the first 1-s interval.

### 13.6 Twin-Tee Notch-Filter Circuit

For the twin-tee notch filter in Fig. P13.6

a) Using admittance values, e.g., $G_1 = 1/R_1$, write Kirchhoff's current law at nodes 2, 3, and 4 to obtain

$$\begin{bmatrix} C_3 s + G_1 + G_2 & 0 & -kC_3 - G_2 \\ 0 & (C_1 + C_2)s + G_3 & -C_2 s - kG_3 \\ -G_2 & -C_2 s & C_2 s + G_2 \end{bmatrix} * \begin{bmatrix} V_2 \\ V_3 \\ V_4 \end{bmatrix} = \begin{bmatrix} G_1 \\ sC_1 \\ 0 \end{bmatrix} V_i$$

Letting $G_1 = G_2 = G_3/2 = G$ and $C_1 = C_2 = C_3/2 = C$ gives

$$\begin{bmatrix} 2(Cs+G) & 0 & -2kCs-G \\ 0 & 2(Cs+G) & -Cs-2kG \\ -G & -Cs & Cs+G \end{bmatrix} * \begin{bmatrix} V_2 \\ V_3 \\ V_4 \end{bmatrix} = \begin{bmatrix} G \\ sC \\ 0 \end{bmatrix} V_i$$

Cramer's rule gives

$$F(s) = \frac{V_3}{V_i} = \frac{\begin{vmatrix} 2G+2sC & 0 & G \\ 0 & 2G+2sC & sC \\ -G & -sC & 0 \end{vmatrix}}{\begin{vmatrix} 2G+2sC & 0 & -G-2ksC \\ 0 & 2G+2sC & -2kG-sC \\ -G & -sC & G+sC \end{vmatrix}} = \frac{s^2 + \omega_o^2}{s^2 + \dfrac{\omega_o}{Q}s + \omega_o^2}$$

where $\omega_o = 1/RC = 1$ krad/s and $Q = [4(1-k)]^{-1}$.

b) The zeros are $s_z = \pm j\omega_o$, and the poles are

$$s_P = \begin{cases} \omega_o\left[-\dfrac{1}{2Q} \pm \sqrt{\left(\dfrac{1}{2Q}\right)^2 - 1}\right] & Q \le 0.5 \\[4mm] \omega_o\left[-\dfrac{1}{2Q} \pm j\sqrt{1-\left(\dfrac{1}{2Q}\right)^2}\right] & Q \ge 0.5 \end{cases}$$

c) With $s = j\omega$

$$F(j\omega) = \frac{\omega_o^2 - \omega^2}{\left(\omega_o^2 - \omega^2\right) + j\dfrac{\omega_o}{Q}\omega}$$

Normalizing frequency $\omega$ by $\omega_o$, and letting $u = \omega/\omega_o$ gives

$$F(ju) = F\left(j\frac{\omega}{\omega_o}\right) = \frac{1-u^2}{\left(1-u^2\right) + j\dfrac{1}{Q}u}$$

Taking the magnitude of F(ju) gives

$$|F(ju)| = \frac{|1-u^2|}{\sqrt{(1-u^2)^2 + \frac{1}{Q^2}u^2}}$$

To solve for the normalized half-power frequency $u_H$, set $|F(ju_H)|^2 = 1/2$, giving

$$\frac{(1-u_H^2)^2}{(1-u_H^2)^2 + \frac{1}{Q^2}u_H^2} = \frac{1}{2}$$

Then

$$u_H^2 \pm \frac{u_H}{Q} - 1 = 0$$

so the normalized half-power frequency is

$$u_H = \sqrt{1 + \left(\frac{1}{2Q}\right)^2} \pm \frac{1}{2Q}$$

The phase function is

$$\angle F(ju) = -a\tan^{-1}\left(\frac{1}{Q}\frac{u}{1-u^2}\right)$$

Therefore, the $\pm 45°$ phase shift normalized frequencies $u_{45}$ are given by

$$u_{45}^2 \pm \frac{u_{45}}{Q} - 1 = 0$$

```
Problem 13.6 - Part d
*       PSpice Simulation of Adjustable-Bandwidth Twin-Tee Filter.
*
.PARAM       twopi={8*atan(1)} k=0
*
Vs     1     0      AC     1
E1     5     0      VALUE = {k*V(3)}
R1     1     2      10k
R2     2     3      10k
R3     4     5      5k
C1     1     4      {100n/twopi}
C2     4     3      {100n/twopi}
C3     2     5      {200n/twopi}
*
.AC     DEC    20     100    10k
.PRINT AC V(3) VP(3)
* The following k values give Q = 0.25, 0.5, 1, and 2
.STEP PARAM k      LIST  0, 0.5, 0.75, 0.875
.PROBE V(3) VP(3)
.END
```

FIGURE 4.7 Netlist to Simulate Twin-Tee Filter

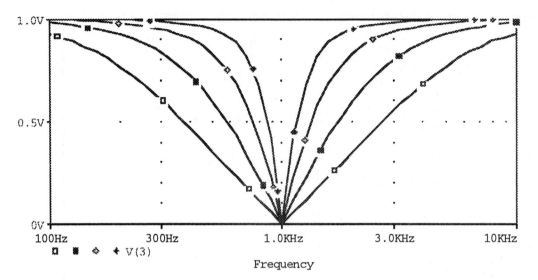

a) Magnitude.

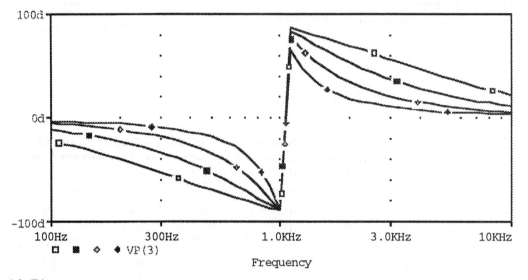

b) Phase.

FIGURE 4.8 Bode Plot

Because this quadratic equation is the same as for the half-power frequency, it is clear that $u_{45}$ and $u_H$ have the same values.

d) To measure the magnitude and phase with PSpice, use the netlist shown in Fig. 4.7. Figures 4.8a and b show the magnitude and phase curves that you can plot with Probe. Now, use Probe's **Cursor** command on the Analog menu and the **Search_commands** item from the Cursor menu to move the cursor to the frequencies where the magnitude equals $1/\sqrt{2}$ V. Set cursor 1 to the left before measuring each curve. Now, measure the half-power frequencies from the Cursor menu using the search command

      search forward for level (0.7071)

```
 Problem 13.6 - Part e
*       PSpice Simulation of Adjustable-Bandwidth Twin-Tee Filter.
*
.PARAM          twopi={8*atan(1)} k=0.875
.OPTIONS DISTRIBUTION GAUSS
Vs      1       0       AC      1
E1      5       0       VALUE = {k*V(3)}
R1      1       2       Rmod    10k
R2      2       3       Rmod    10k
R3      4       5       Rmod    5k
C1      1       4       Cmod    {100n/twopi}
C2      4       3       Cmod    {100n/twopi}
C3      2       5       Cmod    {200n/twopi}
.MODEL          Rmod    RES     (R=1 DEV 10%)
.MODEL          Cmod    CAP     (C=1 DEV 10%)
.AC     DEC     20      100     10k
.PRINT AC V(3) VP(3)
.MC 51 AC V(3)  MIN OUTPUT ALL
.PROBE V(3)
.END
```

FIGURE 4.9 Netlist for Statistical Analysis

twice for each curve. Compare your observations with values that you calculate from the half-power frequency equation above. Repeat this process with the phase shift curve, but search for -45° and 45°, and compare your measurements with calculations of the 45° phase-shift frequencies.

e) The netlist in Fig. 4.9 does the statistical analysis of the twin-tee circuit with ±10% Gaussian variation of each resistance and capacitance. The Monte Carlo command MC has 51 points, because PSpice does the first run with nominal parameter values. To select only the random samples and eliminate the nominal section, choose **Select_sections** from Probe's Section Selection menu. Next, highlight **Select_all_sections** using the arrow keys and then type the spacebar, or click the mouse on **Select_all_sections**. Now, highlight the NOMINAL Monte Carlo run and deselect it by typing the spacebar. These steps leave the remaining 50 sections selected, so type <Return> to proceed to the **Analog** menu.

Using Probe, we obtain the magnitude and phase plots that appear in Fig. 4.10. The magnitude plot in Fig. 4.10a shows that the twin-tee notch-filter magnitude function varies greatly with parameter variations. The phase curve in Fig. 4.10b is surprising. Some of the phase plots show a 180° phase shift at $f_o$, but others indicate that statistical variations cause the phase shift to continue on from -90° to -270°. This effect is due to some zeros, nominally at $\pm j\omega_o$, moving into the right half of the s plane. Those remaining on the imaginary axis or moving into the left half of the s plane have a normal 180° phase discontinuity at $f_o$.

Using the Performance Analysis feature of Probe with the goal functions in Fig. 4.11 gives the histograms shown in Fig. 4.12 a and b. To create these histograms, first select **X_axis** from the Analog menu. Then, at the X_axis menu select **Performance_analysis**. Now, back in the Analog menu, choose **Add_trace**. Respond to the prompt to select a variable by typing

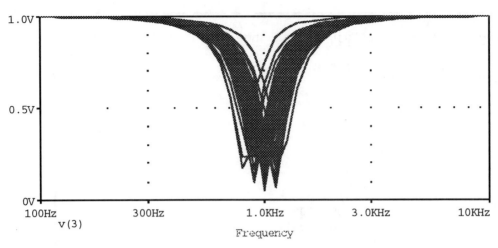

a) Magnitude

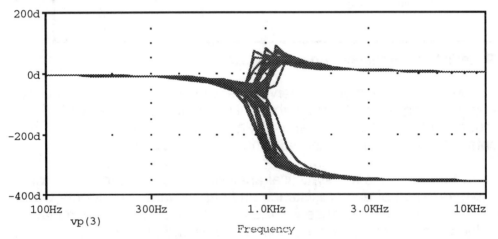

b) Phase

FIGURE 4.10 Monte Carlo Variations

center_freq(v(3)) <Return>

to see the histogram of the center frequency, as shown in Fig. 4.12a. To generate the bandwidth histogram, remove the center frequency histogram using **Remove_trace** on the Analog menu. Next, select **Add_trace** and type

bandwidth(v(3)) <Return>

to see the bandwidth histogram of Fig. 4.12b appear on your screen.

The center-frequency histogram shows that the center frequency remains at about 1 kHz for about 55% of the samples. Of the remaining samples, about 35% occur at 1.1 kHz and 25% at 0.85 kHz. The remaining samples fall into a small group at 0.8 kHz. The bandwidth histogram in Fig. 4.12b shows a more uniform distribution, with about 50% of the samples falling into two groups about the 500-Hz design bandwidth. The center-frequency histogram indicates a standard deviation (sigma) of 85.4 Hz, and the bandwidth histogram shows a sigma of 84.9 Hz. The standard deviation indicates the range of the variable on either side of the median value within which 68% of the samples reside.

```
* Problem 13.6 - Statistical Analysis Goal Functions
half_power(1) = x2-x1
{
     1|
          search for level (0.7071) !1
          search for level (0.7071) !2
     ;
}
bandwidth(1) = x1
{
     1|
          search for min !1
     ;
}
ph10k(1)=y1
{
     1|
          search for xval (9.999k) !1
     ;
}
```

FIGURE 4.11 Goal Functions to Determine Center Frequency and Bandwidth

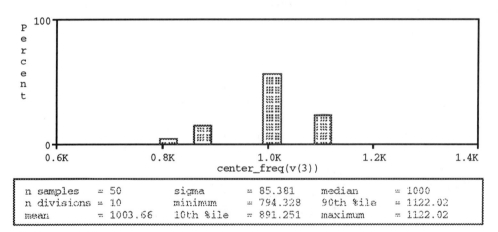

| n samples | = 50 | sigma | = 85.381 | median | = 1000 |
| n divisions | = 10 | minimum | = 794.328 | 90th %ile | = 1122.02 |
| mean | = 1003.66 | 10th %ile | = 891.251 | maximum | = 1122.02 |

a) Center Frequency.

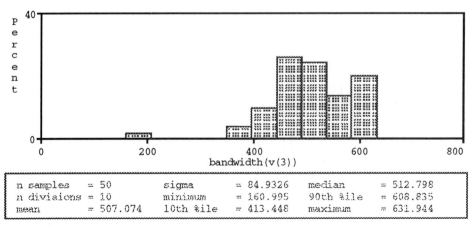

| n samples | = 50 | sigma | = 84.9326 | median | = 512.798 |
| n divisions | = 10 | minimum | = 160.995 | 90th %ile | = 608.835 |
| mean | = 507.074 | 10th %ile | = 413.448 | maximum | = 631.944 |

b) Bandwidth.

FIGURE 4.12 Center Frequency and Bandwidth Statistics of Twin-Tee Circuit

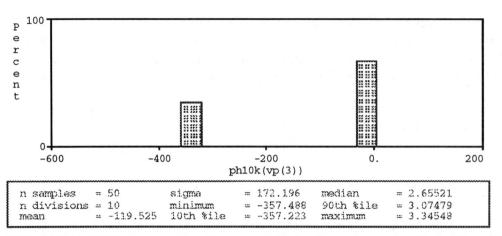

FIGURE 4.13 Histogram for Phase at High Frequency for Twin-Tee Circuit

The histogram of the phase shift at 9.999 kHz, in Fig. 4.13, shows that about 40% of the 50 Monte Carlo simulations have the anomalous behavior, where the phase shift does not have a 180° discontinuity at $f_o$.

## 14.3 Three-Pole Butterworth Transient

For the Laplace-domain circuit shown in Fig. 4.14

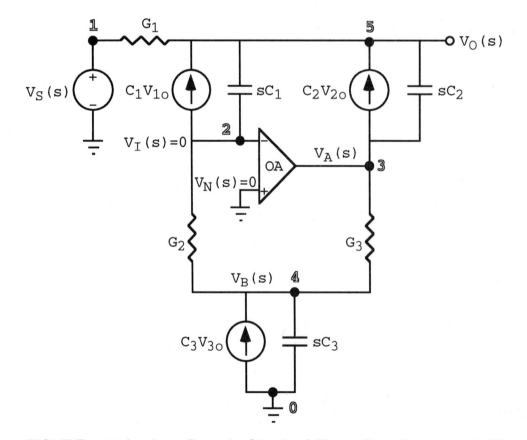

FIGURE 4.14 Laplace-Domain Circuit of Three-Pole Butterworth Filter ($R_1 = 1/G_1 = R_2 = 1/G_2 = R_3 = 1/G_3 = 1/G = 1$ k$\Omega$, $C_1 = C_2 = C_3 = C = 1$ μF)

a) First, find the Laplace solution for the output voltage in order to determine the damping constant(s) for the natural solution. These affect our choice of the transient simulation times <t_Print> and <t_Final>. The matrix form of the three node voltage equations for the Laplace-domain circuit in Fig. 4.14 are

$$
\begin{bmatrix} 0 & -G & -Cs \\ -G & Cs+2G & 0 \\ -Cs & 0 & 2Cs+G \end{bmatrix}
\begin{bmatrix} V_A(s) \\ V_B(s) \\ V_o(s) \end{bmatrix} =
\begin{bmatrix} -CV_{1o} \\ CV_{3o} \\ CV_{1o}+CV_{2o}+GV_s(s) \end{bmatrix}
$$

The determinant $\Delta(s)$ is

$$
\Delta(s) = -C^3\left[ s^3 + 2\frac{G}{C}s^2 + 2\left(\frac{G}{C}\right)^2 s + \left(\frac{G}{C}\right)^3 \right] = -C^3\left( s+\frac{G}{C}\right)\left[ s^2 + \left(\frac{G}{C}\right)s + \left(\frac{G}{C}\right)^2 \right]
$$

$$
= -C^3\left( s+\frac{G}{C}\right)\left( s+\frac{G}{2C} - j\frac{\sqrt{3}}{2}\frac{G}{C}\right)\left( s+\frac{G}{2C} + j\frac{\sqrt{3}}{2}\frac{G}{C}\right)
$$

Using Cramer's rule

$$
V_s(s) = \frac{\begin{vmatrix} 0 & -G & 0 \\ -G & Cs+2G & 0 \\ -Cs & 0 & GV_s(s) \end{vmatrix}}{\Delta(s)} = \frac{(G/C)^3}{s^3 + 2\left(\dfrac{G}{C}\right)s^2 + 2\left(\dfrac{G}{C}\right)^2 s + \left(\dfrac{G}{C}\right)^3} V(s)
$$

With $V_{1o} = V_{2o} = V_{3o} = 0$ and $V_S(s) = 1/s$ and using V, mA, ms, k$\Omega$, and $\mu$F as the units

$$
V_o(s) = \frac{1}{s(s+1)(s^2+s+1)} = \frac{K_1}{s} + \frac{K_2}{s+1} + \frac{K_3}{s+\left(1-j\sqrt{3}\right)/2} + \frac{K_3^{*}}{s+\left(1+j\sqrt{3}\right)/2}
$$

where

$$
K_1 = sV_o(s)\big|_{s=0} = 1\,V
$$

$$
K_2 = (s+1)V_o(s)\big|_{s=-1} = -1\,V
$$

and

$$
K_3 = \left[s+\left(1-j\sqrt{3}\right)/2\right]V_o(s)\bigg|_{s=0} = \frac{1}{s(s+1)\left[s+\left(1+j\sqrt{3}\right)/2\right]}\Bigg|_{s=-(1-j\sqrt{3})/2} = -\frac{1}{2j}\times\frac{2}{\sqrt{3}}\,V
$$

Then

$$
V_o(s) = \frac{1}{s(s+1)(s^2+s+1)} = \frac{1}{s} - \frac{1}{s+1} - \frac{1}{2j}\frac{2/\sqrt{3}}{s+\left(1-j\sqrt{3}\right)/2} + \frac{1}{2j}\frac{2/\sqrt{3}}{s+\left(1+j\sqrt{3}\right)/2}
$$

so the output voltage $v_O(t)$ is

```
Problem 14.3 - Three-Pole Butterworth Transient
*
* Part a) 1 V Unit-Step Source
Vs     1      0      1
*
* Part b) 1 V-ms Unit-Impulse Source
* Vs   1      0      PWL[(0,1000) (2u,0)]
*
* Part c)    exp(-t/2)*cos(sqrt(3)*t/2) V Response
* EVs 1      0      VALUE={exp(-1k*time/2)*cos(sqrt(3)*1k*time/2)}
*
E1     3      0      0      2      100k
R1     1      5      1k
R2     2      4      1k
R3     3      4      1k
C1     5      2      1u
C2     5      3      1u
C3     4      0      1u
.TRAN 0.2m   10m    UIC
.PROBE
.END
```

FIGURE. 4.15 PSpice Netlist for Three-Pole Butterworth Filter

$$v_o(t) = \left[ 1 - e^{-t} - \frac{2}{\sqrt{3}} e^{-t/2} \sin\left( \frac{\sqrt{3}}{2} t \right) \right] u(t)$$

Now, use the netlist in Fig. 4.15 to obtain the Probe plot, in Fig. 4.16. Superimpose the function from above on this plot to see the comparison between the simulation and the theoretical solutions.

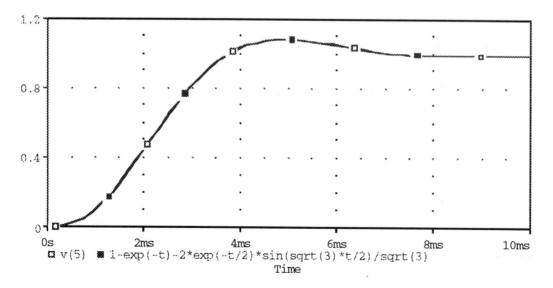

FIGURE 4.16 Probe Plot of PSpice Simulation and Solution

b) When the source waveform is

$$v_s(t) = \delta(t)\ V - ms$$

with time in ms, the Laplace transform $V_O(s)$ is

$$V_0(s) = \frac{1}{(s+1)(s^2+s+1)} = \frac{K_1}{s+1} + \frac{K_2}{s+\frac{1}{2}-j\frac{\sqrt{3}}{2}} + + \frac{K_2^*}{s+\frac{1}{2}+j\frac{\sqrt{3}}{2}}$$

where the residues are

$$K_1 = (s+1)V_o(s)\Big|_{s=-1} = \frac{1}{1-1+1} = 1\ V$$

$$K_2 = \left[s+(1-j\sqrt{3})/2\right]V_o(s)\Big|_{s=-(1-j\sqrt{3})/2} = -\frac{1}{\sqrt{3}}\angle 30°\ V$$

Then, the output voltage is

$$v_o(t) = \left[e^{-t} - \frac{2}{\sqrt{3}}e^{-t/2}\cos\left(\frac{\sqrt{3}}{2}t + 30°\right)\right]u(t)\ V$$

To simulate a 1 V-ms impulse with PSpice requires a short-duration pulse with area equal to 1 V-ms. The PWL transient 1000-V, 2-μs triangle source in the netlist in Fig. 4.15 has a 1 V-ms area. Run this simulation, by removing the asterisk before the PWL source and inserting one before the DC source. Using Probe gives the plot shown in Fig. 4.17. The output voltage calculation also appears in the plot. To simplify entry of the equation, define the macros

```
t  = 1000*time
pi = 4*atan(1)
ph = 30*pi/180
wd = sqrt(3)/2
```

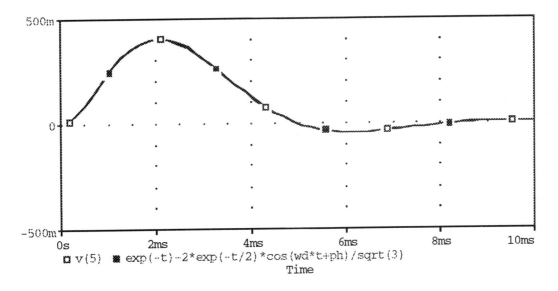

FIGURE 4.17 Response with 1 V–ms Impulse Source

The simulation and the calculation values are the same.

c) When the source voltage waveform is

$$v_S(t) = e^{-t/2} \cos\left(\sqrt{3}t/2\right)u(t) \text{ V}$$

with time in ms, then the Laplace transform $V_O(s)$ is

$$V_o(s) = \frac{s+1/2}{(s+1)(s^2+s+1)^2}$$

$$= \frac{K_1}{s+1} + \frac{K_{21}}{\left[s+(1-j\sqrt{3})/2\right]^2} + \frac{K_{21}^*}{\left[s+(1+j\sqrt{3})/2\right]^2}$$

$$+ \frac{K_{22}}{s+(1-j\sqrt{3})/2} + \frac{K_{22}^*}{s+(1+j\sqrt{3})/2}$$

The residues are

$$K_1 = (s+1)V_o(s)\Big|_{s=-1} = \frac{s+1/2}{(s^2+s+1)^2}\Big|_{s=-1} = -\frac{1}{2}\text{ V}$$

$$K_{21} = \left[s+(1-j\sqrt{3})/2\right]^2 V_o(s)\Big|_{s=(-1+j\sqrt{3})/2} = \frac{s+1/2}{(s+1)\left[s+(1+j\sqrt{3})/2\right]^2}\Big|_{s=(-1+j\sqrt{3})/2}$$

$$= -\frac{1}{2\sqrt{3}} \angle \tan^{-1}\left(\frac{1}{\sqrt{3}}\right)\text{ V/ms}$$

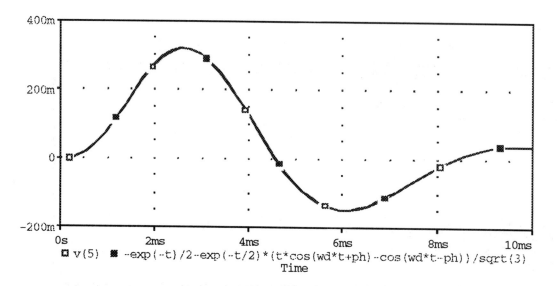

FIGURE 4.18 Exponentially-Damped Cosine-Function Source Response

$$K_{22} = \frac{d}{ds}\left\{\left[s+\left(1-j\sqrt{3}\right)/2\right]^2 V_o(s)\right\}\Bigg|_{s=(-1+j\sqrt{3})/2} = \frac{d}{ds}\left\{\frac{s+1/2}{(s+1)\left[s+\left(1+j\sqrt{3}\right)/2\right]^2}\right\}\Bigg|_{s=(-1+j\sqrt{3})/2}$$

$$= \frac{1}{2\sqrt{3}}\angle -\tan^{-1}\left(\frac{1}{\sqrt{3}}\right)\,\text{V}$$

Then the output voltage is

$$v_O(t) = \left\{-\frac{1}{2}e^{-t} - \frac{e^{-t/2}}{\sqrt{3}}\left[t\cos\left(\frac{\sqrt{3}}{2}t+30°\right) - \cos\left(\frac{\sqrt{3}}{2}t-30°\right)\right]\right\}u(t)\,\text{V}$$

The netlist to simulate the exponentially damped cosine function uses the VALUE form of an E-type source. Run the netlist PR14_3s.CIR, after commenting out the Part a DC source and removing the asterisk in front of the Part c EVs source. Using Probe gives the plot shown in Fig. 4.18. The output voltage waveform from the derivation above falls directly on top of the plot of the simulation waveform. The equation uses the same macros as in Part b to simplify the mathematical form of the expression.

## 15.4 Full-Wave Rectified Sinusoidal Source with LC Filter

The netlist to simulate the circuit shown in Fig. P15.4 appears in Fig. 4.19. While developing the PSpice solution and making calculations for comparison, remember that two frequencies exist for this problem. The first is the 60-Hz frequency $f_{o1}$ of the sinusoid and the second is the 120-Hz frequency of the full-wave rectified sinusoid $f_{o2}$. The netlist uses parameter "twopi" in the VALUE form of the E-type device to define the full-wave rectified sinusoidal source. In this circuit resistance $R_S$ models both source and inductance resistance. The effective coil Q, equal to $\omega_{o2}L/R_S$, equals 121. To calculate the natural frequencies of this circuit, use the characteristic equation

```
Problem 15.4 Full-Wave Rectified Sinusoidal Source with LC Filter
*
.PARAM        twopi={8*atan(1)}
.OPTIONS      NUMDGT=6
E1     1    0      VALUE = {440*sqrt(2)*abs(sin(twopi*60*time))}
RS     1    2      50
RL     3    0      4K
L      2    3      8
C      3    0      33u
*
.TRAN 166.667u 1.0 0.983333 166.667u UIC
.PRINT TRAN I(L) V(1) V(3)
.FOUR 120 5 I(L) V(1) V(3)
.PROBE
.END
```

FIGURE. 4.19 Netlist for Problem 15.4

$$s^2 + \left(\frac{R_s}{L} + \frac{1}{R_L C}\right)s + \left(1 + \frac{R_s}{R_L}\right)\frac{1}{LC} = 0$$

which gives

$$s = -\alpha \pm j\omega_d = -6.913 \pm j61.90 \text{ rad/s}$$

The reciprocal of the damping factor is approximately 145 ms, so the circuit's transient essentially finishes by 1 s. This time is a convenient value for <t_Final>, because the 120th cycle completes in 1 s. To run the simulation for the last two cycles means that <t_No_Print> has to be 0.983333 s. To obtain 50 points for each period, use <t_Print> equal to 166.667 μs. Let <t_Ceil> have this value also so that PSpice cannot try to use the default value of 20 ms. In the FOUR statement, remember to use the 120-Hz frequency of the full-wave rectified waveform. Since the function of the LC filter is to suppress the AC fluctuations, we do not expect the waveforms to be rich in harmonics. For this reason, the FOUR command limits the number of harmonics to five.

The source and output voltage simulation waveforms appear in Fig. 4.20. Figure 4.21 shows the output waveform using a smaller vertical scale, in order to see the slight variation of the output voltage. The simulation shows that a steady-state level of operation does not yet exist. The current waveform in Fig. 4.22 shows that the current never reduces to zero.

Figure 4.23 shows the running average of the instantaneous power that the source supplies and the load receives. The values these have after each period equals the periodic average, so the cursor values in Fig. 15.4.4 show that the source power equals 38.466 W, and the load power is 37.958 W. The efficiency is

$$\eta = \frac{P_L}{P_s} \times 100\% = \frac{37.958}{38.466} \times 100 = 98.7\%$$

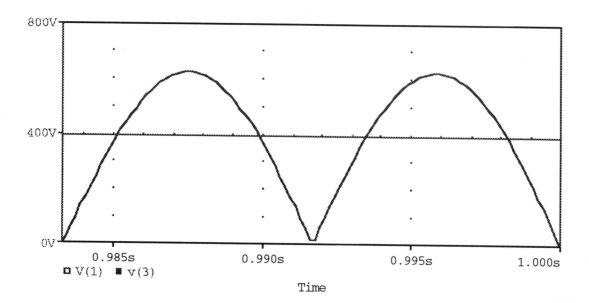

FIGURE. 4.20 Voltage waveforms $v_S(t)$ and $v_O(t)$

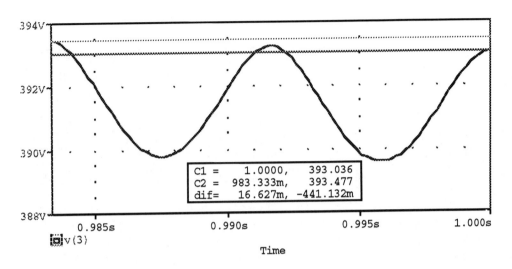

FIGURE. 4.21 Voltage $v_O(t)$ Waveform

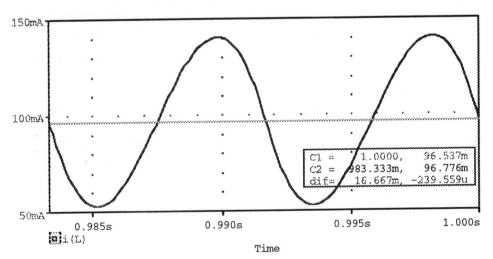

FIGURE. 4.22 Current $i_L(t)$ Waveform

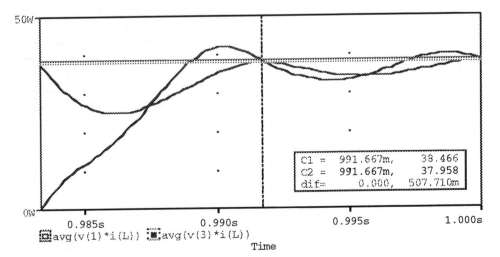

FIGURE. 4.23 Running Average of $v_S(t)i_L(t)$ and $v_O(t)i_L(t)$

DC COMPONENT =    9.696343E-02

| HARMONIC NO | FREQUENCY (HZ) | FOURIER COMPONENT | NORMALIZED COMPONENT | PHASE (DEG) | NORMALIZED PHASE (DEG) |
|---|---|---|---|---|---|
| 1 | 1.200E+02 | 4.397E-02 | 1.000E+00 | -1.795E+02 | 0.000E+00 |
| 2 | 2.400E+02 | 4.351E-03 | 9.895E-02 | -1.798E+02 | -2.372E-01 |
| 3 | 3.600E+02 | 1.231E-03 | 2.800E-02 | -1.798E+02 | -2.371E-01 |
| 4 | 4.800E+02 | 5.073E-04 | 1.154E-02 | -1.796E+02 | -4.342E-02 |
| 5 | 6.000E+02 | 2.552E-04 | 5.803E-03 | -1.791E+02 | 4.461E-01 |

TOTAL HARMONIC DISTORTION =   1.036432E+01 PERCENT

Table 4.1 Fourier Components of $i_L(t)$

DC COMPONENT =    3.959784E+02

| HARMONIC NO | FREQUENCY (HZ) | FOURIER COMPONENT | NORMALIZED COMPONENT | PHASE (DEG) | NORMALIZED PHASE (DEG) |
|---|---|---|---|---|---|
| 1 | 1.200E+02 | 2.640E+02 | 1.000E+00 | -9.000E+01 | 0.000E+00 |
| 2 | 2.400E+02 | 5.285E+01 | 2.002E-01 | -9.001E+01 | -8.023E-03 |
| 3 | 3.600E+02 | 2.268E+01 | 8.590E-02 | -9.003E+01 | -2.899E-02 |
| 4 | 4.800E+02 | 1.263E+01 | 4.782E-02 | -9.007E+01 | -6.569E-02 |
| 5 | 6.000E+02 | 8.059E+00 | 3.052E-02 | -9.012E+01 | -1.171E-01 |

TOTAL HARMONIC DISTORTION =   2.250911E+01 PERCENT

Table 4.2 Fourier Components of $v_1(t)$

DC COMPONENT =    3.912640E+02

| HARMONIC NO | FREQUENCY (HZ) | FOURIER COMPONENT | NORMALIZED COMPONENT | PHASE (DEG) | NORMALIZED PHASE (DEG) |
|---|---|---|---|---|---|
| 1 | 1.200E+02 | 1.768E+00 | 1.000E+00 | 8.882E+01 | 0.000E+00 |
| 2 | 2.400E+02 | 9.575E-02 | 5.415E-02 | 6.956E+01 | -1.926E+01 |
| 3 | 3.600E+02 | 2.952E-02 | 1.670E-02 | 3.982E+01 | -4.901E+01 |
| 4 | 4.800E+02 | 1.862E-02 | 1.053E-02 | 2.391E+01 | -6.492E+01 |
| 5 | 6.000E+02 | 1.433E-02 | 8.106E-03 | 1.838E+01 | -7.044E+01 |

TOTAL HARMONIC DISTORTION =   5.820754E+00 PERCENT

Table 4.3 Fourier Components of $v_O(t)$

## Simulation Fourier Coefficients

The Fourier coefficients that PSpice finds for the current $i_L(t)$, source voltage $v_1(t)$, and output voltage $v_O(t)$ appear in Tables 4.1 through 4.3, respectively.

## Fourier Coefficient Calculations

The voltage $v_S(t)$ has the full-wave form

$$v_s(t) = V_{sm}\left|\sin\omega_{o1}t\right|$$

$$= \frac{2V_{sm}}{\pi}\left\{1 + \sum_{k=1}^{\infty}\frac{2}{4k^2 - 1}\sin(2k\omega_{o1}t - 90°)\right\}$$

The sinusoidal form of the Fourier series allows easy comparison with the PSpice results shown above, which express Fourier series coefficients in sine-series form. The series admittance $Y(s)$ relates the phasor current I to the phasor source voltage $V_S$ and is

$$Y(s) = \frac{I}{V_s} = \frac{1}{sL + R_s + R_L \parallel (1/sC)}$$

$$= Y_o \frac{\tau_{C1}s + 1}{\tau_{L1}\tau_{C2}s^2 + (\tau_{L2} + \tau_{C2})s + 1}$$

where

$$\tau_{L1} = \frac{L}{R_s}$$

$$\tau_{L2} = \frac{L}{R_s + R_L}$$

$$\tau_{C1} = R_L C$$

$$\tau_{C2} = (R_s \parallel R_L)C$$

and

$$Y_o = 1/(R_s + R_L)$$

The voltage divider theorem gives the transfer function $T(s)$, relating phasor voltage $V_o$ to the phasor source voltage $V_S$

$$T(s) = \frac{V_o}{V_s} = \frac{R_L \parallel (1/sC)}{sL + R_s + R_L \parallel (1/sC)}$$

$$= \frac{T_o}{\tau_{L1}\tau_{C2}s^2 + (\tau_{L2} + \tau_{C2})s + 1}$$

where

$$T_o = \frac{R_L}{R_s + R_L}$$

Using $s = j\omega_k$ in the transfer function $Y(s)$, shows that the kth harmonic phasor $I_k$ of the current is

| k | $\lvert V_{sk} \rvert$, V | $\angle V_{sk}$, ° | $\lvert V_{ok} \rvert$, V | $\angle V_{ok}$, ° | $\lvert I_k \rvert$, A | $\angle I_k$, ° |
|---|---|---|---|---|---|---|
| 1 | 264.09 | -90 | 1.771 | 91.06 | 4.408E-02 | -179.52 |
| 2 | 52.82 | -90 | 0.088 | 90.53 | 4.386E-03 | -179.76 |
| 3 | 22.64 | -90 | 0.017 | 90.35 | 1.252E-03 | -179.84 |
| 4 | 12.58 | -90 | 0.005 | 90.26 | 5.214E-04 | -179.88 |
| 5 | 8.00 | -90 | 0.002 | 90.21 | 2.654E-04 | -179.90 |

Table 4.4 Calculated Fourier Coefficients $V_{sk}$, $V_{ok}$, and $I_k$

$$I_k = Y_o \frac{1 + j\omega_k \tau_{C1}}{\left(1 - \omega_k^2 \tau_{L1}\tau_{C2}\right) + j\omega_k\left(\tau_{L2} + \tau_{C2}\right)} V_{sk}$$

$$= Y_o \frac{1 + jc}{a + jb} V_{sk}$$

with

$$a = 1 - \omega_k^2 \tau_{L1}\tau_{C2}$$

$$b = \omega_k\left(\tau_{L2} + \tau_{C2}\right)$$

$$c = \omega_k \tau_{C1}$$

From the transfer function T(s), the kth harmonic phasor $V_{ok}$ of the output voltage is

$$V_{ok} = \frac{T_o}{\left(1 - \omega_k^2 \tau_{L1}\tau_{C2}\right) + j\omega_k\left(\tau_{L2} + \tau_{C2}\right)} V_{sk} = \frac{T_o}{a + jb} V_{sk}$$

The kth harmonic phasor $V_{sk}$ of the full-wave source $v_s(t)$ is

$$V_{sk} = \frac{4V_{sm}}{\pi} \frac{1}{4k^2 - 1} \angle -90°$$

Using these equations and a spreadsheet program gives the solution in Table 4.4, where $V_s$ = 440 V RMS, $f_{o1}$ = 60 Hz, $R_S$ = 50 Ω, $R_L$ = 4 kΩ, L = 8 H, and C = 33 μF. The magnitudes from the spreadsheet program and the PSpice magnitudes are in close agreement. The PSpice phase values agree well with the calculations, except for the output voltage phase for k greater than 1. Here, the small size of the voltage component at the second and higher harmonic frequencies gives PSpice a numerically difficult task.

The power delivered to the load equals the power dissipated in load resistance $R_L$

$$P_{Load} = \frac{V_{o(rms)}^2}{R_L} = \frac{V_o^2}{R_L} + \sum_{k=1}^{\infty} \frac{\lvert V_{ok} \rvert^2}{2R_L}$$

The power absorbed by the source resistance $R_S$ is

$$P_{RS} = R_S I_{(rms)}^2 = R_S I_o^2 + \frac{1}{2}\sum_{k=1}^{\infty} R_S \lvert I_k \rvert^2$$

Summing the power dissipation in the source resistance and the load power gives the power that the source supplies

$$P_{Source} = P_{Load} + P_{RS}$$

The spreadsheet calculations in Table 4.5 that use these equations show that the power supplied by the source equals 38.797 W while the power that the load receives is 38.269 W, so the power conversion efficiency is

$$\eta = \frac{P_L}{P_s} \times 100 = \frac{38.269}{38.797} \times 100 = 98.6\,\%$$

The PSpice values for the source and load power are each slightly less than the calculation values. This difference is, in part, due to the PSpice solution not yet being in periodic steady-state operation.

| k | $P_L$, W | $P_{RS}$, W | $P_S$, W |
|---|---|---|---|
| 0 | 38.269 | 0.478 | 38.747 |
| 1 | 0.0 | 0.049 | 0.049 |
| 2 | 0.0 | 0.0 | 0.0 |
| 3 | 0.0 | 0.0 | 0.0 |
| 4 | 0.0 | 0.0 | 0.0 |
| 5 | 0.0 | 0.0 | 0.0 |
| Total = | 38.269 | | 38.797 |

Table 4.5 Average Power Values

## 17.3 Coefficient of Coupling for Maximum Power

a) To simulate the circuit shown in Fig. P17.3, write the netlist shown in Fig. 4.24. The netlist uses E- and G-type sources to model the ideal transformer. Using the VALUE form of these lets the STEP command sweep the turns ratio n. The PARAM statement gives n a value of 2. This value has no importance, because the OPT command suppresses the bias solution. For convenience, we give parameter wo a value equal to $2\pi \times 60$ and use this value to define the inductance values. To simplify the average power calculation, the magnitude of the AC source $V_S$ has the RMS value of 440 V.

b) To calculate the average power use the formula

$$P = Re\{VI^*\}$$

When you run Probe using PR17_3s.DAT, after selecting the sections for n equal to 2,4,6, and 8 from the Section selection menu, select the Macros menu and enter the macro functions

    conj(x)=m(x)*m(x)/x

and

    avgpwr(v,i)=r(v*conj(i))

Exit the Macros menu and use the Add_trace menu to enter

```
Problem 17.3 - Adjusting Ideal Transformer Turns Ratio for Maximum Power
*
.PARAM        wo={8*atan(1)*60} n=2
.OPT   NUMDGT=6        NOBIAS
*
Vs      1      0      AC      440
RS      1      2      1
RL      4      5      80
LS      2      3      {4/wo}
LL      5      0      {60/wo}
*      The ideal transformer
Eideal         40     0      VALUE={n*V(3)}
Vms            4      40
Gideal         0      3      VALUE={n*I(Vms)}
*
.AC    DEC    20     6      600
.STEP LIN     PARAM n      2      8      0.2
.PROBE
.END
```

FIGURE 4.24 PSpice Netlist for Problem 17.3

avgpwr(v(3),i(rs))

Alternately, with a word processor you can write these macro functions before you run Probe. Save this MAC file with the name PR17_3s.MAC. Be sure to store this file in the subdirectory or folder that contains the other Problem 17 files. This file already exists on the EXAMPLES disk in the Ch17 subdirectory or folder. Using an IBM PC or compatible computer, start Probe by typing

PROBE /M PR17_3s.MAC PR17_3s.DAT

With a Macintosh, the macro becomes available if the macro file has the same name as the DAT file, has the MAC suffix, and exists in the DAT file's folder. The plot that you see shows that at 60 Hz, the average power maximizes for a value of the turns ratio n near 5. Adjust the horizontal  scale to show the frequency range from 10 to 100 Hz. Use the **Label** item on the Analog menu to mark the curves with the values of n. The Probe plot appears as in Fig. 4.25. Set the cursor to the 60-Hz frequency using the **Cursor** item of the Analog menu and the **Search_commands** item of the Cursor menu. Enter

sxv(60)

to move the cursor to the 60-Hz frequency. Now click on each of the data markers below the horizontal axis to change the cursor attachment from the curve for n equal to 2 to each of the other curves. Since the 60-Hz frequency is not among the actual data, you will have to reset the cursor to 60 Hz each time using either "sfxv(60)" or "sbxv(60)." Watch the values of the average power to see that a maximum occurs for n between 4 and 6.

c) To plot the average power as a function of the turns ratio n, we use the **Performance_analysis** item of the X_axis menu. This feature of Probe requires that we define a goal function to measure responses at 60 Hz. The

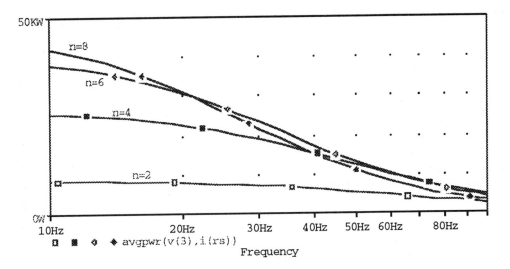

FIGURE. 4.25 Average Power vs. Frequency with Turns Ratio as Parameter

goal function in Fig. 4.26 moves the cursor to 60 Hz, marks this point, and returns the y-coordinate value of this point. This goal function exists as the file PR17_3s.GF on the EXAMPLES disk. To have this goal function and the macros from PR17_3s.MAC available using an IBM PC or compatible computer, start Probe by typing

PROBE /M PR17_3s.MAC /G PR17_3s.GF PR17_3s.DAT

With a Macintosh, the macro becomes available if the macro file has the same name as the DAT file, has the GF suffix, and exists in the DAT file's folder. From the Section Selection menu choose **All_Ac_sweep**. Next, select **X_axis** from the Analog menu and **Performance_analysis** from the X_axis menu. Now you see the Analog menu again and the plot's horizontal axis has become turns ratio n. Select **Add_trace** and enter

Val60(avgpwr(v(3),i(rs)))

to plot the average power. Select the **Cursor** item from the Analog menu and **maX** from the Cursor menu to find the maximum value. The plot now appears as in Fig. 4.27 The cursor information box shows that the maximum power is 10.572 kW and occurs for n equal to 5.0.

```
Val60(1)=y1
{
 1|
    search forward xval(60) !1

 ;
}
```

FIGURE. 4.26 Goal Function to Measure Values at 60 Hz

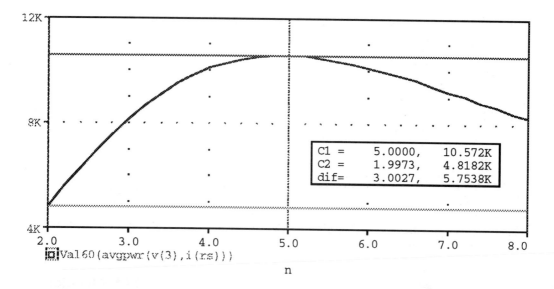

FIGURE. 4.27 Average Power vs. Frequency with Turns Ratio as Parameter

## Theory

For the circuit shown in Fig. P17.3, the source impedance at 60 Hz is

$$Z_s = R_s + jX_s = R_s + j\omega L_s = 1 + j4 \; \Omega$$

and the load impedance is

$$Z_L = R_L + jX_L = R_L + j\omega L_L = 80 + j60 \; \Omega$$

The power into the ideal transformer maximizes when the magnitude of the turns ratio n makes the magnitude of the ideal transformer's input impedance equal to the magnitude of the source impedance. This situation requires that

$$n = \sqrt[4]{\frac{R_L^{\,2} + X_L^{\,2}}{R_s^{\,2} + X_s^{\,2}}} = \sqrt[4]{\frac{80^2 + 60^2}{1^2 + 4^2}} = 4.925$$

Therefore, the average power is

$$P = \mathrm{Re}\{V_3 I_{RS}^{\,*}\} = \mathrm{Re}\left\{ \left( \frac{Z_L / n^2}{Z_s + Z_L / n^2} V_s \right) \left( \frac{V_s}{Z_s + Z_L / n^2} \right)^{*} \right\}$$

$$= \frac{|V_s|^2}{n^2} \frac{\mathrm{Re}\{Z_L\}}{\left| Z_s + Z_L / n^2 \right|^2} = \frac{440^2}{100/\sqrt{17}} \times \frac{80}{\left| 1 + j4 + \dfrac{80 + j60}{100/\sqrt{17}} \right|^2} = 10575 \; \text{W}$$

# Appendix A PSpice and Probe Syntax

## PSpice and Probe Scaling Suffixes

| F | femto | $10^{-15}$ |
|---|-------|-----------|
| P | pico | $10^{-12}$ |
| N | nano | $10^{-9}$ |
| U | micro | $10^{-6}$ |
| M | milli | $10^{-3}$ |
| K | kilo | $10^{3}$ |
| MEG | Mega | $10^{6}$ |
| G | Giga | $10^{9}$ |
| T | Tera | $10^{12}$ |

Table A.1 SPICE Scaling Factors (Probe uses m for milli and M for Mega)

## PSpice Expressions

| Function | Expression | Comment |
|----------|-----------|---------|
| abs(x) | $\lvert x \rvert$ | |
| exp(x) | exp(x) | |
| sin(x) | sin(x) | x in radians |
| cos(x) | cos(x) | x in radians |
| tan(x) | tan(x) | x in radians |
| atan(x)=arctan(x) | $\tan^{-1}(x)$ | result in radians |
| log(x) | ln(x) | log base e |
| log10(x) | log(x) | log base 10 |
| pwr(x,y) | $\lvert x \rvert^y$ | |
| pwrs(x,y) | $\lvert x \rvert^y$ (x > 0), $-\lvert x \rvert^y$ (x < 0) | |
| sqrt(x) | $\sqrt{x}$ | |
| table(x,$x_1$,$y_1$,..., $x_n$,$y_n$) | | $y_1$ for x<$x_1$, $y_n$ for x>$x_n$ inerpolates for $x_1$<x<$x_2$ |
| limit(x, min, max) | | min for x<min, max for x>max, otherwise x |

Table A.2 PSpice Expressions

## PSpice  Commands

.AC LIN/OCT/DEC <Points> <Start_Freq> <End_Freq>

.DC [LIN] <Sweep_Var> <Start_Val> <End_Val> <Inc>
+      [Nested_Sweep]

.DC DEC/OCT <Sweep_Var> <Start_Val> <End_Val> <Points>
+      [Nested_Sweep]

.DC  <Sweep_Var> LIST <Val_1> <Val_2> ...
+      [Nested_Sweep]

.DISTRIBUTION    <Name>      (<Dev_1>, <Prob_1>, ...)

.END

.ENDS
        Ends a subcircuit description (See .SUBCKT).

.FOUR <Frequency> [<No_Harmonics>] <Output_1> ...

.FUNC <Fct_Name>([<Arg_1>, ...]) <Fct_Def>

.IC V(<Node_1>  [, <Node_2>]) = <Value_1> ...

.INC "<File_Name>"

.LIB ["<File_Name>"]

.LOADBIAS "<File_Name>"

.MC <No_Runs> <Analysis> <Output_1> ...
+      <Function> [<Option> ...] [SEED = <Value>]; See  Sec.  2.5.5

.MODEL <Model_Name> [AKO: <Ref_Model_Name>] <Type_Name>
+          [([<Param_1>=<Value_1> [<Tolerance_Spec>]] ...
+          [T_MEASURED=<T_Measured>]
+          [T_ABS=<T_Abs>/T_REL_GLOBAL=<T_Rel_Global>/
+          [T_REL_LOCAL=<T_Rel_Local>])]

        <Tolerance_Spec> =
            [DEV <value>[%][<L_&_D>]] [LOT <value>[%][<T_&_D>]]

        <L_&_D> =
            /<Lot_No> /<Distribution_Name>

| Type Name | Character | Device |
|---|---|---|
| RES | R | Resistance |
| CAP | C | Capacitance |
| IND | L | Inductance |
| D | D | Diode |
| NPN | Q | npn bipolar junction transistor (BJT) |
| PNP (LPNP) | Q | (lateral) pnp bipolar junction transistor |
| NJF | J | n-channel junction field-effect transistor (JFET) |
| PJF | J | p-channel junction field-effect transistor |
| NMOS | M | n-channel metal-oxide semiconductor field-effect transistor (MOSFET) |
| PMOS | M | p-channel metal-oxide semiconductor field-effect transistor |
| GASFET | G | n-channel GaAs field-effect transistor (MESFET) |
| CORE | K | nonlinear magnetic core (transformer) |
| VSWITCH | S | voltage-controlled switch |
| ISWITCH | W | current-controlled switch |

Table A.3 <Type_Name> and Character Symbols for MODEL Statement

.NODESET V(<Node_1>[, <Node_2>]) = <Value_1> ...

.NOISE V(<Node_1> [, <Node_2>]) <Source> [Interval]

.OP
 Gives operating point information, including linearization parameters of nonlinear devices, in the output file.

.OPTIONS [<Option_Name>] ... [<Option_Name> = <Value>] ...

| Option | Meaning |
|---|---|
| ACCT | Calls for accounting information to appear |
| EXPAND | Shows subcircuit expansions |
| LIBRARY | Lists lines from libraries |
| LIST | Lists circuit elements |
| NOBIAS | Suppresses node-voltage bias values |
| NODE | Lists the node table |
| NOECHO | Suppresses listing of the input file |
| NOMOD | Suppresses listing of model parameters and temperature changes |
| NOOUTMSG | Suppresses error messages in output file |
| NOPAGE | Suppresses paging and banner production |
| NOPRBMSG | Suppresses error messages in DAT file |
| NOREUSE | Suppresses reuse of bias point |
| OPTS | Lists values for options |

Table A.4 Flag Options

| Option | Meaning | Unit | Default |
|--------|---------|------|---------|
| ABSTOL | Accuracy of currents | A | 1 p |
| CHGTOL | Accuracy of charge | C | 10 f |
| CPTIME | CPU time for run | s | 1 M |
| DISTRIBUTION | Default Monte Carlo distribution | | UNIFORM |
| GMIN | Minimum branch conductance | S | 1 f |
| ITL1 | DC and bias blind iteration limit | | 40 |
| ITL2 | DC and bias educated guess iteration limit | | 20 |
| ITL4 | Transient iteration point limit | | 10 |
| ITL5 | Total transient iteration limit (ITL5=0 means ∞) | | ∞ |
| LIMPTS | Maximum print or plot points (LIMPTS=0 means ∞) | | ∞ |
| NUMDGT | Significant digits (< 8) | | 4 |
| PIVREL† | Relative pivot magnitude | | 1 m |
| PIVTOL† | Absolute pivot magnitude | | 100 a |
| RELTOL | Relative V and I accuracy | | 1 m |
| TNOM | Nominal temperature | °C | 27 |
| VNTOL | Voltage accuracy | V | 1 μ |
| WIDTH | Same as .WIDTH OUT = 80/132 | | 80 |

Table A.5 Value Options († Change these at your own risk)

.PARAM <Name> = <Value>/{<Expression>} ...

.PLOT DC/AC/NOISE/TRAN
+       <Output_Var1> [(<Lower_Limit>, <Upper_Limit>)] ...

.PRINT DC/AC/NOISE/TRAN <Output_1> ...

| Modifier | Effect |
|----------|--------|
| M | Magnitude (The default) |
| P | Phase |
| R | Real part |
| I | Imaginary part |
| DB | 20 times log of value |
| G | Time delay ($=-\partial\phi/\partial\omega$) |

Table A.6 Modifier Letters

.PROBE [<Output_1> ...]

```
.SAVEBIAS "<File_Name>" DC/OP/TRAN [NOSUBCKT]
+     [TIME = <Value> [REPEAT]]
+     [TEMP = <Value>]
+     [STEP = <Value>]
+     [MCRUN = <Value.]
+     [DC = <Value> / DC1 = <Value_1.> [DC2 = <Value_2>]]


.SENS <Output_1> ...


.STEP [LIN] <Sweep_Var> <Start_Val> <End_Val> <Inc>
+     [Nested_Sweep]

.STEP OCT <Sweep_Var> <Start_Val> <End_Val> <Points>
+     [Nested_Sweep]

.STEP DEC <Sweep_Var> <Start_Val> <End_Val> <Points>
+     [Nested_Sweep]

.STEP <Sweep_Var> LIST <Val_1> <Val_2> ...
+     [Nested_Sweep]


.SUBCKT <Subckt_Name> <Node_1> <Node_2> ... <Node_n>
+     [OPTIONAL: <Interface_Node1> = <Default_1> ...]
+     [PARAMS: <Parameter_Name_1> = <Value_1> ...]
...
...
.ENDS


.TEMP <Temperature_Value_1> ...


.TF <Output> <Input_Source>


.TRAN [/OP] <t_Print> <t_Final> [<t_No_Print> [<t_Ceil> ]] [UIC]
```

| Parameter | Meaning | Units | Default |
|---|---|---|---|
| <t_Print> | Time between print values | s | |
| <t_Final> | Final time of simulation. | s | |
| <t_No_Print> | Time before printing | s | 0 |
| <t_Ceil> | Ceiling on numerical time steps | s | <t_Final>/50 |

Table A.7 Transient Specification Parameters

.WATCH DC/AC/TRAN <Output_Variable> [(<Lower-Limit>,
+       <Upper_Limit>)] ...


.WCASE <Analysis> <Output_Var> <Function>
+       [<Option_1> ...]


.WIDTH OUT = 80/132
       Default is 80 (Also, see WIDTH option for .OPTIONS command).


## PSpice Devices


C<name> <Node_1> <Node_2> [<Model_Name>] <Value>
+       [IC=<IC_Value>]

| Parameter | | Units | Default |
|---|---|---|---|
| C | Capacitance multiplier | | 1 |
| VC1 | Linear voltage coefficient | $V^{-1}$ | 0 |
| VC2 | Quadratic voltage coefficient | $V^{-2}$ | 0 |
| TC1 | Linear temperature coefficient | $C^{-1}$ | 0 |
| TC2 | Quadratic temperature coefficient | $C^{-2}$ | 0 |

Table A.8 Capacitance Model Parameters


E<Name> <Node_1> <Node_2> <Node_3> <Node_4> <Value>

E<Name> <Node_1> <Node_2>
+       POLY(<Dim>)(<Node_3>, <Node_4>) ... <Coefficients>

E<Name> <Node_1> <Node_2> VALUE = {<Expression>}

E<Name> <Node_1> <Node_2> TABLE {<Expression>} =
+       = (<Input_1>, <Output_1>), ... , (<Input_n>, <Output_n>)

E<Name> <Node_1> <Node_2> FREQ {<Expression>} =
+       (<Freq_1>, <Mag_1>, <Phase_1>), ... ,
+       (<Freq_n>, <Mag_n>, <Phase_n>)

E<Name> <Node_1> <Node_2> LAPLACE {<Expression>} =
+       {<Laplace_Transform>}

F<Name> <Node_1> <Node_2> V<Control> <Val>

F<Name> <Node_1> <Node_2>
+     POLY(<Dim>)(<Node_3>, <Node_4>) ... <Coefficients>


G<Name> <Node_1> <Node_2> <Node_3> <Node_4> <Val>

G<Name> <Node_1> <Node_2>
+     POLY(<Dim>)(<Node_3>, <Node_4>) ... <Coefficients>

G<Name> <Node_1> <Node_2> VALUE = {<Expression>}

G<Name> <Node_1> <Node_2> TABLE {<Expression>} =
+     = (<Input_1>, <Output_1>), ... , (<Input_n>, <Output_n>)

G<Name> <Node_1> <Node_2> FREQ {<Expression>} =
+     (<Freq_1>, <Mag_1>, <Phase_1>), ... ,
+     (<Freq_n>, <Mag_n>, <Phase_n>)

G<Name> <Node_1> <Node_2> LAPLACE {<Expression>} =
+     {<Laplace_Transform>}


H<Name> <Node_1> <Node_2> V<Control> <Val>

H<Name> <Node_1> <Node_2>
+     POLY(<Dim>)(<Node_3>, <Node_4>) ... <Coefficients>


I<Name> <Node_1> <Node_2> [DC] [<DC_Value>]
+     [AC <AC_Mag> <AC_Phase>]
+     <Transient_Specification>

      <Transient_Specification> =
            $EXP(<V_1> <V_2> <t_{d1}> <\tau_1> <t_{d2}> <\tau_2>)$
            $PULSE(<V_1> <V_2> <t_d> <t_r> <t_f> <t_p> <t_{per}>)$
            $PWL(<t_1> <V_1> <t_2> <V_2> ... )$
            $SFFM(<V_1> <V_2> <f_c> <M> <f_m>)$
            $SIN(<V_1> <V_2> <f> <t_d> <\alpha> <\phi>)$


K<Name> <L_Name1> <L_Name2> ... <Value>


L<name> <Node_1> <Node_2> [<Model_Name>] <Val>
+     [IC=<IC_Val>]

| Parameter | | Units | Default |
|-----------|--|-------|---------|
| L | Inductance multiplier | | 1 |
| IL1 | Linear voltage coefficient | $A^{-1}$ | 0 |
| IL2 | Quadratic voltage coefficient | $A^{-2}$ | 0 |
| TC1 | Linear temperature coefficient | $C^{-1}$ | 0 |
| TC2 | Quadratic temperature coefficient | $C^{-2}$ | 0 |

Table A.9 Inductance Model Parameters

R<name> <Node_1> <Node_2> [<Model_Name>] <Val>
+       [TC=<TC1>, [<TC2>]]

| Parameter | | Units | Default |
|-----------|--|-------|---------|
| R | Resistance multiplier | | 1 |
| TC1 | Linear temperature coefficient | $°C^{-1}$ | 0 |
| TC2 | Quadratic temperature coefficient | $°C^{-2}$ | 0 |
| TCE | Exponential temperature coefficient | $\%/°C^{-1}$ | 0 |

Table A.10 Resistance Model Parameters

S<Name> <Plus_Node> <Minus_Node>
+       <Plus_Control> <Minus_Control> <Model_Name>

| Parameter | Meaning | Units | Default |
|-----------|---------|-------|---------|
| RON | On resistance | $\Omega$ | 1.0 |
| ROFF | Off resistance | $\Omega$ | $10^6$ |
| VON | On control voltage | V | 1.0 |
| VOFF | Off control voltage | V | 0 |

Table A.11 VSWITCH Parameters with Default Values

V<Name> <Node_1> <Node_2> [DC] [<DC_Value>]
+       [AC <AC_Mag> <AC_Phase>]
+       <Transient_Specification>

    <Transient_Specification> =
        EXP(<$V_1$> <$V_2$> <$t_{d1}$> <$\tau_1$> <$t_{d2}$> <$\tau_2$>)
        PULSE(<$V_1$> <$V_2$> <$t_d$> <$t_r$> <$t_f$> <$t_p$> <$t_{per}$>)
        PWL(<$t_1$> <$V_1$> <$t_2$> <$V_2$> ... )
        SFFM(<$V_1$> <$V_2$> <$f_c$> <M> <$f_m$>)
        SIN(<$V_1$> <$V_2$> <f> <$t_d$> <$\alpha$> <$\phi$>)

W<Name> <Plus_Node> <Minus_Node> <Control_V_Source>
+       <Model_Name>

| Parameter | Meaning | Units | Default |
|-----------|---------|-------|---------|
| RON | On resistance | $\Omega$ | 1.0 |
| ROFF | Off resistance | $\Omega$ | $10^6$ |
| ION | On control voltage | A | 1.0 m |
| IOFF | Off control voltage | A | 0 |

Table A.12 ISWITCH Parameters with Default Values

X<Name> <Node_1> <Node_2> ... <Node_n> <Subckt_Name>
+     [PARAMS: <Par_Name_1 > = <Call_Value_1> ...]

## Probe Information and Syntax

<Macro_Name>[(<Arg_1>,...)]=<Definition>

<Goal_Function_Name>(1,[2,...,n,<Sub_Arg1>,...]) =
     <Marked_Point_Expression>
{
 1|
    <Search_Commands_and_Marked_Points_for_1>
 ;
 [2|
    <Search_Commands_and_Marked_Points_for_2>
 ;
 etc.]
}

Search [<direction>] [/<Start_Point>/] [#<Consecutive_Points>#]
[(<Range_x>)[,(<Range_y>)]] [for] [<Repeat>:] <Condition>]; See Sec. 2.6.5

| Option | Suffix | Default Name | File Function |
|--------|--------|--------------|---------------|
| /C | CMD | | A command file to run Probe |
| /D | DEV | PSPICE.DEV | Defines display and hard-copy devices |
| /G | GF | PROBE.GF | Defines goal functions for performance analysis |
| /L | LOG | | Records actions that you can replay |
| /M | MAC | PROBE.MAC | Defines macro functions |
| /S | DSP | PROBE.DSP | Allows display of previous sessions |

Table A.13 Probe Files

| Function | Expression | Comment |
|---|---|---|
| sgn(x) | 1 x>0, -1 x<0, 0 x=0 | |
| m(x) | Magnitude of x | x may be complex |
| p(x) | Phase of x | In degrees |
| r(x) | Real part of x | |
| img(x) | Imaginary part of x | |
| g(x) | Group delay of x | |
| d(x) | Derivative of x | With respect to X_axis |
| s(x) | Integral of x | With respect to X_axis |
| avg(x) | Running average of x | With respect to X_axis |
| avgx(x, d) | Running average of x from x-d to x | Over range of X_axis |
| rms(x) | Running RMS average of x | Over range of X_axis |
| db(x) | Magnitude of x in decibels | |
| min(x) | Minimum of real part of x | |
| max(x) | Maximum of real part of x | |

Table A.14 Additional Probe Functions

# Index